OUTDOOR SCIENCE

A PRACTICAL GUIDE

STEVE RICH

OUTDOOR SCIENCE

A PRACTICAL GUIDE

National Science Teachers Association

Arlington, Virginia

National Science Teachers Association

Claire Reinburg, Director
Jennifer Horak, Managing Editor
Andrew Cocke, Senior Editor
Judy Cusick, Senior Editor
Wendy Rubin, Associate Editor
Amy America, Book Acquisitions Coordinator

ART AND DESIGN
Will Thomas Jr., Director
Joe Butera, Senior Graphic Designer, cover and interior design
Cover photos by Simon McConico, Jens Stolt Photography, Linda Bucklin, Lisa Thornberg, and Konradlew for iStock

PRINTING AND PRODUCTION
Catherine Lorrain, Director

SCILINKS
Tyson Brown, Director
Virginie L. Chokouanga, Customer Service and Database Coordinator

NATIONAL SCIENCE TEACHERS ASSOCIATION
Francis Q. Eberle, PhD, Executive Director
David Beacom, Publisher

Library of Congress Cataloging-in-Publication Data
Rich, Steve A.
  Outdoor science : a practical guide / by Steve A. Rich.
    p. cm.
  Includes bibliographical references and index.
  ISBN 978-1-935155-12-6
  1. Science--Study and teaching--Activity programs. 2. Nature study--Study and teaching--Activity programs. 3. Outdoor education. I. Title.
  Q181.R535 2010
  508.071--dc22
                    2009046466

eISBN 978-1-936137-78-7

NSTA is committed to publishing material that promotes the best in inquiry-based science education. However, conditions of actual use may vary, and the safety procedures and practices described in this book are intended to serve only as a guide. Additional precautionary measures may be required. NSTA and the authors do not warrant or represent that the procedures and practices in this book meet any safety code or standard of federal, state, or local regulations. NSTA and the authors disclaim any liability for personal injury or damage to property arising out of or relating to the use of this book, including any of the recommendations, instructions, or materials contained therein.

Featuring sciLINKS®—a new way of connecting text and the internet. Up-to-the minute online content, classroom ideas, and other materials are just a click away. For more information, go to www.scilinks.org/Faq.aspx.

CONTENTS

Dedication

Outdoor Science: A Practical Guide is dedicated with love to my son, Spencer Anthony Rich, who as a young boy helped me rediscover the outdoors through his eyes, and as a young man continues to help me see the world in new ways.

As Maya Angelou said, "My son is my monument."

Foreword

Science teaching offers a realm of possibilities for creating new kinds of classrooms and learning communities that fit the needs of today's students. Following the 2005 release of Richard Louv's *Last Child in the Woods* and the No Child Left Inside national initiative, more attention has been focused on the need for increased opportunities for children to interact with nature during the school day. Louv coined the term *nature-deficit disorder* to describe children's disconnections from the natural environment as a result of spending more time in front of televisions and video games. This disconnection has led to serious, troubling trends among today's schoolchildren: a rise in obesity, a lack of empathy for living things, more attention disorders, and social isolation, among other concerns. Furthermore, in addition to health and behavioral problems, serious learning problems result when students do not have opportunities early on and throughout their educations—both in school and during play—to interact with the outdoor world.

Many of us recall a childhood in which we spent most of the day playing outdoors. Whether it was the backyard, woods, fields, ponds, or streams in our rural or suburban neighborhoods or inner-city parks, playgrounds, harbors, empty lots, sidewalks, or alleys—we could

explore, interact with, and come to know our natural world through play while simultaneously building social skills. Digging in the soil, watching ants, catching frogs, collecting seeds, floating sticks, making "dams and rivers," skipping rocks, observing butterflies and bees—all of these experiences contribute to children's early understanding of the living and physical worlds. Sadly, the "outside school" lives of today's children are much more structured and supervised, with few opportunities to interact in outdoor play spaces. When children do have free time, it often is spent in front of a computer, the television, or video games. Likewise, the school day is spent primarily indoors with few opportunities to interact in a naturalized outdoor setting. The once-a-year field trip may be the only time some children ever experience learning in an outdoor setting.

But there is a way to not only bring nature to our students but also bring our students to nature. This book offers a practical, effective solution for dealing with the nature-deficit disorder epidemic by creating a special kind of classroom environment and collaborative community that makes it possible for every child to participate in outdoor exploration and discovery that bring science and other subjects to life. As today's students prepare to be the adults of tomorrow's society, we need to transform the traditional four-wall, "contrived" learning environment into one in which students can participate in practical projects and activities that are not restricted by the physical boundaries of the classroom or limitations of bringing the outside world inside.

Outdoor Science: A Practical Guide describes how teachers can change the physical environment for learning to an outdoor classroom where students can feel successful as they interact with the natural world; in addition, Steve Rich presents innovative ideas for transforming curriculum and instruction to meet the needs of today's learners. The suggestions in this powerful book are consistent with the recommendations from the National Science Foundation's report "Environmental Science and Engineering for the 21st Century," which confirms the importance of environmental education for building knowledge, critical- and creative-thinking skills, and basic life skills such as problem solving, consensus building, information management, and strong verbal and written communication skills (NSF 2000). The outdoor classroom offers an ideal setting in which students can delve into important environmental issues and develop an appreciation of their natural surroundings, which offer an authentic vehicle for creating an enhanced sense of stewardship and appreciation for the natural world.

Our nation's future depends on educating the next generation of students to be wise and responsible stewards of the environment. To prepare students, we need to design effective instruction and develop innovative rigorous curriculum, and offer new types of learning spaces that provide opportunities for student-driven, inquiry-based, and interdisciplinary learning that values school-community partnerships. Now you too can learn how to engage your students, school, and community in designing the types of classrooms that the author has developed successfully during his accomplished teaching career. From simple outdoor spaces to extensive habitat areas, every school can find a way to design an outdoor learning space using the practical suggestions in this comprehensive guide.

Thank you, Steve, for sharing your tremendous wisdom, insights on science teaching and learning, and innovative and practical ideas for moving education outside, where children can truly engage in authentic learning in natural surroundings. In an era of standards and accountability, we need to think outside the box for new ways to make learning accessible to students. What better way to do that than to take "the box" outside!

Page Keeley
Maine Mathematics and Science Alliance
NSTA President 2008–09

Correlation to National Science Education Standards

	Content Standards	Chapter 3 Activities	Chapter 4 Activities	Chapter 5 Activities	Chapter 6 Activities
Physical Science	K–4 Properties of objects and materials		p. 62		p. 124
	5–8 Properties and changes of properties in matter				
	K–4 Position and motion of objects		p. 84		
	5–8 Motions and forces				
	K–4 Light, heat, and magnetism		p. 78		
	5–8 Transfer of energy		p. 78		
Life Science	K–4 Characteristics of organisms			p. 90	
	5–8 Structure and function in living systems	pp. 44, 46, 48		p. 98	
	K–4 Life cycles of organisms	pp. 42, 50	p. 68	pp. 90, 94, 102, 106	
	5–8 Reproduction and heredity		p. 70		
	K–4 Organisms and environments	pp. 32, 52	p. 72	pp. 96, 98	
	5–8 Regulation and behavior	p. 38	p. 64		
	5–8 Populations and ecosystems	p. 54		p. 100	
	5–8 Diversity and adaptations of organisms	p. 34			
Earth and Space Science	K–4 Properties of Earth materials				p. 124
	5–8 Structure of the Earth system				
	K–4 Objects in the sky				pp. 116, 120
	5–8 Earth's history				p. 122
	K–4 Changes in Earth and sky		pp. 74, 80	p. 92	
	5–8 Earth in the solar system				
Science and Technology	K–4 Abilities to distinguish natural objects/manmade	p. 52			p. 128
	K–4/5–8 Abilities of technological design		p. 84		
	K–4/5–8 Understanding about science and technology				
Science in Personal and Social Perspectives	K–4/5–8 Personal health				
	K–4 Characteristics and changes in populations	pp. 32, 36, 40		p. 100	pp. 130, 132
	5–8 Populations, resources, and environments	pp. 32, 36, 40	p. 76	p. 100	
	K–4 Types of resources			p. 104	
	5–8 Natural hazard				
	K–4 Changes in environments			p. 104	p. 126
	5–8 Risks and benefits			pp. 108, 110	
	K–4/5–8 Science and technology in local challenges/society				
History and Nature of Science	K–4/5–8 Science as a human endeavor	p. 56	p. 66		pp. 114, 116, 122, 126
	5–8 Nature of science				
	5–8 History of science				

Creating a Space for Learning

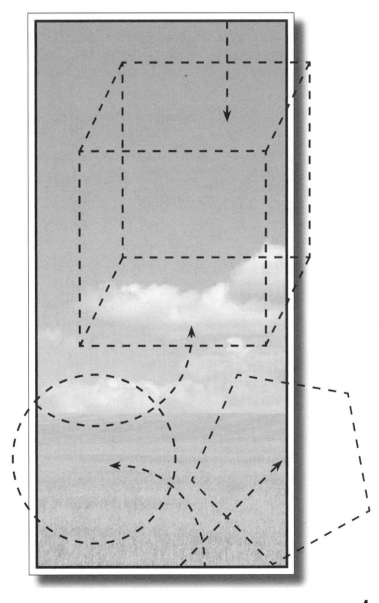

Do you ever look out the window on a nice day and dream about what fun it would be to hold class outside? Your students probably feel the same way! There is a great solution for you and your students: Create an outdoor learning laboratory that can be a classroom center for science, math, language arts, and social studies lessons.

In recent years, No Child Left Behind legislation has forced many educators to focus on math and reading. A subsequent legislative effort, No Child Left Inside, gives credence to the relevance of outdoor learning spaces—formal outdoor classrooms or otherwise. Multiple reports from around the nation, provided by the State Education and Environment Roundtable, show that environment-centered education improves student achievement, builds community partnerships, and even provides an effective context for learning mathematics. Whatever your school's setting—urban, suburban, or rural—there are ways to create an exciting outdoor classroom so your students can have this experience.

Teaching outdoors is rewarding for both teachers and students, and there are many options for creating an outdoor learning laboratory. If you are ambitious and have the resources, you can create an outdoor classroom to use from year to year, building and adding to it whenever you are able. You can also base great lessons on temporary outdoor learning solutions. In this book, I offer suggestions for both alternatives.

How Does This Fit in With My Curriculum?

To discover how effective outdoor classes can be, look at objectives and standards in your curriculum in the areas of life, Earth, and environmental sciences. You can fulfill many of those objectives using simple, inexpensive outdoor lessons—without a field-trip permission slip in sight! In this book I will offer math, social studies, and language arts activities to tie into your curriculum as well.

Many outdoor classroom components that I will address were designed for sixth graders originally but can be adapted easily for elementary school students. Here are some tips to fit outdoor lessons into your curriculum effectively:

Primary Grades (PreK–2)

As you begin the school year in the early grades, think about the concepts that your students need to learn. Some common topics in early standards are sorting and classification, life cycles of plants and animals, objects in the sky, and living versus nonliving things. You may need to adapt the activities in this book for lower grades, but here are suggested lessons for each of these concepts or topics.

Sorting/Classification:
Hunting for Numbers (p. 62)

For young students, either make a simple similar chart or do not use a chart at all. Assign pairs of students items to find, such as two leaves or one rock.

Life Cycles of Plants and Animals:
Monarch Butterfly Life Cycle (p. 42)

This is essentially a labeling activity that you can pair with having a live caterpillar or butterfly in the classroom.

Objects in the Sky:
It's About Time and Human Sundial
(p. 116)

You can do this lesson without any permanent space. Some state standards include making shadows, so this lesson may serve a dual purpose.

Living Versus Nonliving Things:
What Can You Learn From a Seed?
(p. 68)

You can adapt this lesson and use the handout as your own guide rather than as a student handout. Begin the activity by giving each student a seed and asking if the seed is dead or alive.

Upper Elementary Grades (3–5)

Students in these grades are ready to develop more of the abilities and understanding they need to inquire about science topics introduced in previous grades. In particular, the National Science Education Standards address the subject of organisms and their environments with the following explanation: "Humans change environments in ways that can be beneficial or detrimental for themselves or other organisms" (NRC 1996, p. 129). As students mature in upper elementary grades, these lessons from *Outdoor Science* may advance their conceptual understanding of humans' effects on the environment.

Organisms and Their Environments:
Animal Habitat Survey (p. 32)

How do the school building and the people who use it affect animals in the school yard?

Solving an Ecological Mystery (p. 108)

What effect does human activity have on the organisms in a community?

The Characteristics of Organisms: Animals Living on the School Grounds (p. 52)

What characteristics best equip animals to survive in areas that have been developed by humans?

Middle Grades (6–8)

For science students in middle grades, standards and instruction tend to be more specialized, classified as Earth, life, and physical sciences. *Outdoor Science* has lessons that are suited to each of these areas. Here are two samples for each area.

Earth Science: Measuring and Analyzing Erosion (p. 80)

For middle school students, this activity may take on a community service aspect. Students and teachers can survey school grounds for areas of erosion, and students may research, suggest, and test various remedies for the erosion problem.

Weather or Not (p. 74)

This activity will prove useful for middle schoolers if you make it more complex. Teachers can have students take additional weather measurements. If the teacher is trained in GLOBE (*www. globe.gov*), students can gather data to upload to the web, contributing to a pool of scientific data gathered around the world.

Life Science: Graphing Animal Behavior (p. 64)

Students in middle school are developmentally ready to analyze animal behavior. You may even want to come up with other animals in your school yard that have observable behaviors to graph. For example, are there ants that build mounds in a sidewalk crack over and over? Teachers may guide students through selecting an animal behavior to observe and research.

Animal "Arti-Fact" or Fiction? (p. 98)

You may want to consult the language arts teachers at your grade level to make this an integrated learning experience for your students. The science teacher can guide students through finding artifacts and discussions of what evidence the artifact provides. Then the student may make inferences, and the language arts teacher can provide support for writing the fictional story based on fact.

Physical Science: Simple Machines Are for the Birds (p. 84)

Take this lesson further for middle school students by asking them to combine two or more simple machines to make a complex machine that will fill a bird feeder.

What's Your Net Worth? (p. 66)

Extend this lesson beyond what is there for middle school students. Relate it to simple machines. Is the net's handle a lever? Is the arm using the net a lever?

Getting Started

Your first challenge is to decide whether you want to use temporary solutions (see p. 14) or lay the groundwork for a permanent outdoor learning environment. Either way, you should consider the steps in this chapter to help you evaluate your direction.

For a successful outdoor program in which you can create a more permanent space, you will need a good garden plan; student, parent, faculty, and administrator involvement; and financial support.

Setting Up a Good Plan

Begin by identifying the resources you already have and those you will need. The School Yard Inventory chart on page 4 will help you do this. Walk around your school's grounds as you complete the form. You may be surprised when you discover how many resources you already have in place.

Next, take a look at some plans that have been successfully developed at other schools (see pp. 6–11). As you look at the plans, pay attention to the components that grab your attention. Which components fit into both your curriculum and your school's site? Which ones could you accomplish with the resources you presently have available? Which ones will let

Table 1.1. School Yard Inventory Chart

Resource	Available	Feasible	Not Feasible	Estimated Costs
Garden plot				
Mature trees				
Young trees				
Flower containers				
Flower beds				
Wooded area for nature trail				
Shaded area				
Group seating				
Area safe from traffic				
Birdbath				
Butterfly nectar plants				
Butterfly host plants				
Pond or creek				
Patio/asphalt area for chalk projects				
Access to water/ hose				

you fulfill your curricular objectives through classes that are fascinating for you (because your enthusiasm is critical to student enthusiasm) and to your students?

Use the elements that you identify in your screening process as the best ones for your plan. Combine them to draw up your own preliminary plans and get ready to share your vision!

Getting Students, Faculty, and Parents Involved

After you have prepared the inventory and begun planning, give your project a head start on success: Show your ideas and information to your school principal and other administrators. You may wish to speak to your students, the faculty, and the PTA or other parent groups, too. The more ownership your stakeholders feel, the more they will work toward your outdoor classroom's success.

Before you start developing the space, work with your principal to get approval of the outdoor classroom location from your school system. In worst-case scenarios, there have been school systems that removed outdoor classroom plants, benches, or other materials without notice because of rules in place regarding the grounds. It certainly would be difficult to watch a grounds crew haul away materials that you may have purchased with money from grants or fund-raisers. Be proactive so you can avoid this altogether.

Another way to raise the ownership level is to take students on a second inventory survey over the school grounds. Students may well be your best advisers in planning the outdoor learning lab; if they help plan the outdoor learning lab, they will want to help maintain it as well. You may also consider holding a student contest for outdoor

classroom designs, then trying to create the images needed to help students visualize these ideas. A student or an involved parent who is a good artist might sketch a plan for your school yard vision. An art teacher is another great resource; he or she may draw a plan for you or help a student with the plan as part of an art project.

Take all ideas into consideration; then start to fine-tune your plan. Base this tuning on the resources your school yard already has, resources it needs, and funding you expect to receive.

Financing the Outdoor Classroom

Think that you will never find the money to finance your outdoor learning lab dream? Think again! There is more help available than you might guess. You just need to think creatively. Here are just a few "hidden" resources:

- Master gardeners from the local cooperative extension service are often willing to help plan and establish school gardens.

- A landscape architect or garden professional might offer an in-kind donation. Parents also might have skills and business resources to donate.

- The owner of a paving company could donate sand for an archaeological dig area (see p. 13).

- A parent or colleague with plants in his or her own garden could share some perennials to create a butterfly garden (see p. 12).

- A parent who is a restaurant manager or chef might partner with the school on an herb garden (see p. 12).

You might also consider seeking grant money. As nonprofits, schools are eligible for grants. Schools also benefit frequently from the donations of charitable individuals and corporations. In Chapter 2, you will find resources and reproducible sample letters to start your funding search. The chapter also offers suggestions for navigating the fund-raising and grant-writing processes.

When your plan is ready, make a final presentation to your principal. Remember to follow proper procedures for approval—many plans have been rejected simply because someone decided not to go through the necessary channels. Have all plans signed and dated by the proper authorities, and be sure to discuss maintenance procedures before beginning your work. This step ensures that you have the resources in place to not only build but also maintain your outdoor classroom.

School Yard Inventory

Take this inventory when you are outdoors on your school's grounds (Table 1.1, p. 4). Do not try to complete it from memory because you may miss valuable resources. Extra lines have been included so that you can add any unique resources that your school grounds offer.

Outdoor Classroom Plans, Big and Small

An outdoor learning environment can be as small as one math patio (see p. 7) bracketed by two small garden plots. It can be as large as space, resources, and imagination permit. On pages 8 through 12, you will find plans for successful outdoor classrooms that have been developed at other elementary and mid-dle schools. You will also find descriptions of patios and different types of gardens in this chapter. Your own outdoor learning laboratory can include ideas from these plans that you customize for your own outdoor classroom.

Materials

The plans you will see on the following pages use many different plants and the following materials. I have included tips about installation to help you consider which materials you may want to use and what resources you will need to begin the building process.

Landscape timbers with non-arsenic preservatives are made of treated lumber, which will last for years outdoors. To secure the timbers to the ground, drill a hole near each end of a timber with a ¾ in. drill bit. Drive an 8 or 10 in. spike through the timber and into the ground.

Patio blocks come in several colors. Chalk is easiest to see on red or terra-cotta blocks. If you consider another color, buy a sample block and experiment with chalk before making a large purchase.

Before you lay the patio, you will need to level the ground with heavy equipment brought in by a business partner or volunteer. You can also use the manpower of students, parents, and teachers working together. Use a leveling tool to check the area. You can spread a layer of fine gravel used specifically for the final leveling process, although a patio can be built successfully without it. Putting down dark plastic or weed cloth before installing the blocks will inhibit weed growth through any cracks between patio blocks.

Pathways through outdoor learning spaces are essential for protecting plants and other resources. Stepping stones surrounded by non-floating cypress mulch will keep pathways neat and attractive. Avoid having pebble pathways because some students may be tempted to throw the pebbles.

Wooden benches are a good choice for seating. If wooden benches fall over, they will be less likely than concrete benches to injure children. Plus, they can be built inexpensively. (If you are not a carpenter, seek a parent or colleague for help building the benches.) Treated timber with non-arsenic preservatives often comes in 2 in. × 12 in. × 8 ft. boards that can be cut in half to be used as bench tops. Treated posts can be put into the ground for legs. The finished seating will stand up to student use and the elements and can be repaired with relative ease.

Math Patio Options

A math patio is a central element of an outdoor learning laboratory. The math patio can provide a visual way to teach challenging math concepts by allowing students with different learning styles to see a concrete representation of an idea. A math patio can be created for a few hundred dollars, but you cannot put a price tag on the value the math patio brings to your lessons. Here are different types of math patios that you can consider for inclusion in your plan.

A hundreds chart patio is a 10 ft. × 10 ft. grid of concrete blocks (1 ft. square) that allows students to create hundreds charts, graphs, and scale drawings. In the pattern at the top of the page, the numbers of the hundreds chart have been started. Each square represents a concrete block, and students write the numbers with chalk. Although this patio is the larger of the two basic types, you will have more working space if your plan includes only one patio.

Note: This patio can also be used as a calendar math patio if you mark the desired

1	2	3	4	5	6	7	8	9	10

number of days in a month with chalk. Because each row will have 10 squares, however, the calendar will not look like the standard month representation, which might confuse some students.

A calendar math patio is made of square concrete blocks in a 7 ft. × 6 ft. grid to create a standard calendar. You can double the numbers in the last row for the last day or two of the month, a common practice on paper calendars. A calendar math patio will be less than half as expensive as a hundreds patio because it uses far fewer blocks.

S	M	T	W	TH	F	S
					1	2
3	4	5	6	7	8	9
10	11	12	13	14	15	16
17	18	19	20	21	22	23
24/31	25	26	27	28	29	30

In the diagram on the previous page, each square represents a concrete block. The letters and numbers would be written by your students. Notice that this set of blocks is laid out in a 7 ft. x 6 ft. grid. You need an extra row across the top for the days of the week, and you can have five rows for weeks (four would work if you need to save money or space).

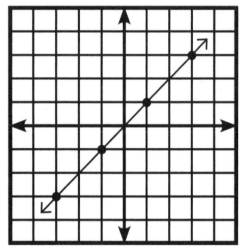

A math decicenter or graphing patio is a variation of the previous math patios (see above). You can build it using larger concrete blocks than you would use in the elementary model. Students can use this center when plotting ordered pairs of numbers on a coordinate graph, using chalk to mark and label the grid and locations of points. Line and bar graphs can also be created on the math patio.

Outdoor Classroom Plans
Elementary Grades
Plan A:
Calendar Patio and Two Gardens
An outdoor classroom with one calendar patio and two compact garden plots is well suited for a small area or a limited budget (see graphic at top of next column). This plan can easily be expanded

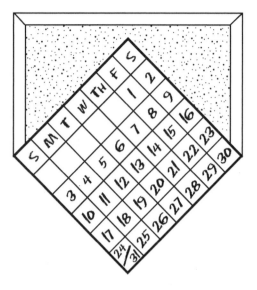

later as the following plans show. You may find that one patio bordered with two garden areas is enough dedicated space for your school and outdoor teaching plans. In the plans described, landscape timbers surround the small gardens.

Even a small garden space can be enhanced with plants to attract butterflies, a birdbath to invite birds, or various plants centered around a theme. (See pp. 12–13 for ideas that will help you select features for this simple permanent installation.)

Plan B:
Two Patios and Six Garden Areas
If you have success with Plan A, think about a future expansion of your outdoor classroom space. Plan B offers twice as much space as Plan A, with six triangles for gardens (see graphic below). The plan is also flexible.

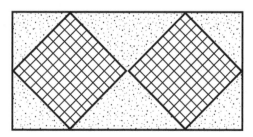

Triangular areas can be turned into gardens or sitting areas. Patios can be created from two calendar grids, two hundreds charts, or one of each. If your school has kindergarten through grade 5 classes, you can suggest that each grade level take responsibility for establishing and maintaining one of the garden areas. Let students plant the gardens as they wish, or you can discuss with them the various garden ideas presented on pages 12 and 13.

Plan C: Comprehensive Plan for Elementary School Outdoor Classroom

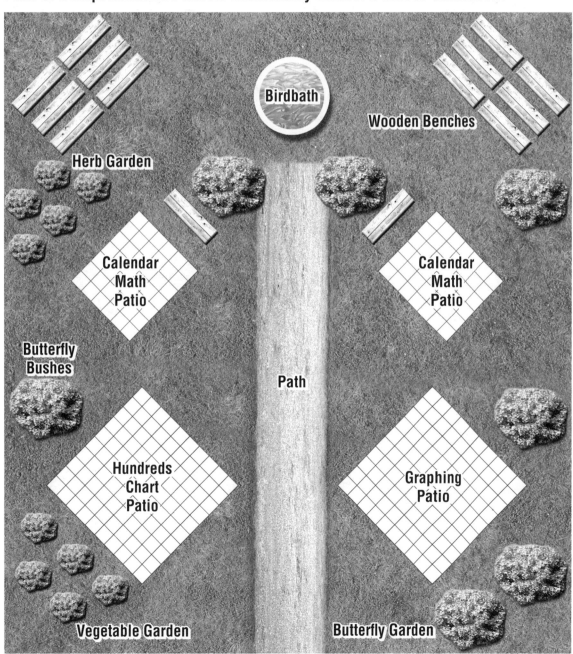

Birdbath

Wooden Benches

Herb Garden

Calendar Math Patio

Calendar Math Patio

Butterfly Bushes

Path

Hundreds Chart Patio

Graphing Patio

Vegetable Garden

Butterfly Garden

Middle Grades

Middle school students have different learning needs than elementary students; your outdoor classroom learning lab design for these students should reflect their needs. For example, you can consider using larger concrete blocks for your middle school patio than you would for an elementary school patio.

Different-color blocks can help define various work areas for small groups. When planning the garden spaces, choose areas that you think would be most useful to your school and curriculum. (For ideas, see the suggestions beginning on p. 12.)

Plan A: Garden and Two-Color Patio

This simple version of a middle school plan uses the multicolor patio blocks described earlier and combines them with one of the theme gardens described on pages 12 through 13.

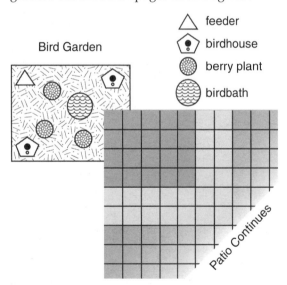

Bird Garden

△ feeder
⬠ birdhouse
◯ berry plant
◯ birdbath

Patio Continues

Plan B: Patio and Four Gardens

Another successful plan incorporates four of the garden–activity center ideas from pages 12 and 13. The gardens are grouped around the decicenter or math patio. The patio can serve as

the central teaching area, and the gardens frame the learning lab, providing space to study plants, birds, and insects. The archaeological dig area provides great science–social studies crossover opportunities. This design also addresses middle school students' needs by offering layered possibilities for in-depth lessons. The design shown provides a perfect separation of activities to support small-group work.

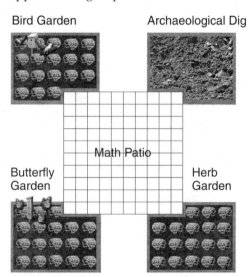

Bird Garden   Archaeological Dig

Math Patio

Butterfly   Herb
Garden   Garden

Comprehensive Plans for Elementary and Middle School Outdoor Classrooms

Plan C: It's Your Choice

You have many options for interesting gardens and activity areas. What you select should depend on your curriculum, resources, and budget. With these factors in mind, tailor your plan to include the gardens of your choice. Any of the triangular garden areas in the school plans on pages 8 through 10 can be turned into one of the gardens listed on pages 12 and 13. You may also want to include garden walkways made with stepping stones or patio blocks to make it easier for students to maintain their outdoor learning classroom. In addition to these ideas, you should brainstorm

Plan C: Comprehensive Plan for Middle School Outdoor Classroom

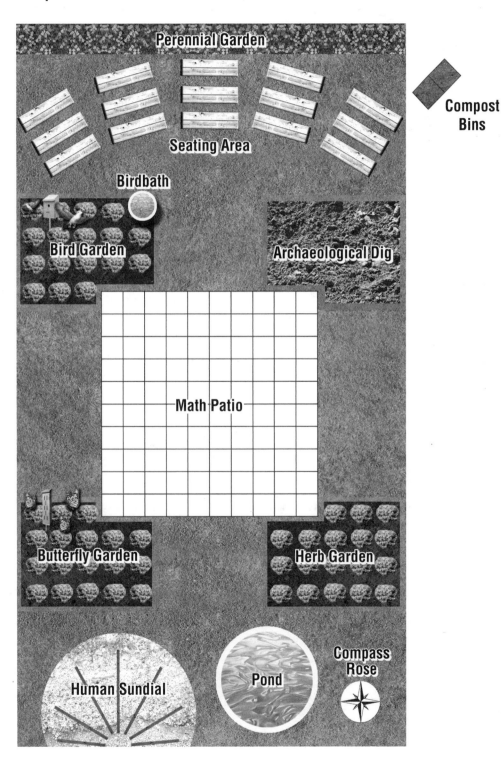

with students so that they have input into the final design. Of course, what you plant will be determined by the garden zone in which you live. Be sure to check gardening books for information about specific plants to determine whether they will flourish in your geographic zone.

Different Kinds of Gardens

Bird Garden

A bird garden attracts birds to your outdoor classroom and should include food, water, shelter, and a place to raise young birds. Shelter might be a small tree that your students plant. Food can come from feeders and natural sources such as berry-yielding plants (blueberries, blackberries) or native grapevines. Purchase a birdbath, or make one by turning a flowerpot upside-down and fastening a tray to the top. Complete the garden by installing a few birdhouses. Your students will be fascinated as they watch birds lay eggs and raise their young.

Butterfly Garden

Butterflies will visit gardens that provide nectar, caterpillar host-plants, and sunshine. The common butterfly bush, buddleia, is hardy and will produce flowers with the nectar that butterflies prefer. Most flowers will attract butterflies, but some of their favorites are lantana, marigolds, zinnias, purple coneflowers, and vinca. Host plants are specific to butterfly species, and females will choose those plants for laying their eggs. Many varieties of swallowtails prefer cold-hardy parsley, which will grow outdoors for much of the school year in most regions of the United States. Other swallowtail host plants are dill and fennel. Monarch butterflies lay eggs on milkweed. Several varieties of milkweed are native to Canada

but grow throughout the United States and as far south as Mexico. Painted lady butterflies prefer thistle weed. Try these plants in sunny areas, and watch the butterflies come!

Herb Garden

An herb garden provides many learning opportunities. Parsley, sage, rosemary, thyme, dill, and mint are easy-to-grow staples. Parsley will provide a place for butterflies to lay eggs, as will dill. Students will find that herbs have distinct aromas. Try having students close their eyes and rub various herbs on their hands, then identify the herbs by their scents. Dill will remind students of pickles, and mint will remind them of gum or candy. Tell students that kitchen gardens in colonial America were planted with a variety of herbs that were used to season and preserve food, mix medicines, create toiletries, and even make housekeeping products. Have students research the roles that herbs played in early American households.

Vegetable Garden

Vegetable gardens can be used in many ways. Here are just a few ideas:

- Kindergarten students studying traditions can plant vegetables as soon as school starts and then use the vegetables they harvest in their own Thanksgiving feasts.

- Math students can use information on seed packets to determine the amount of space each plant needs to grow and how many seeds they need to fill up their gardens.

- Students can practice measuring skills by planting seeds at the correct distances apart.

- Students can plant pumpkin seeds and predict the size of the pumpkins!

Native American Garden

Students can explore Native American culture firsthand as they plant corn, squash, and beans to create a "three sisters" garden. Native American tribes planted one seed from each plant in the same hills. This created an ecosystem with a natural fertilizing effect. This practice also provided a series of harvests instead of just one. Crops that complement each other are still used in practices such as crop rotation. In addition to the science used in this system of companion planting, there are many legends connected to the "three sisters" that you can use for language arts connections. Another option for a Native American garden would be to plant pole beans on rods that lean in to form a teepee under which students can sit. More ideas may be found at the U.S. Bureau of Indian Affairs website (*www.doi.gov/bureau-indian-affairs.html*).

Perennial Garden

Perennial plants regenerate year after year, so a perennial garden is a sound financial investment. Perennials provide great science study and connect with other subjects as well—for example, art teachers at some schools have requested perennials as subjects for their students to draw. Perennials either come back from the same roots and stem each spring or do not completely die during the winter. Grasses such as purple fountain grass and zebra grass are attractive additions to any outdoor classroom. A number of varieties of nandina keep their leaves throughout the school year and provide vibrant colors seasonally. Other perennials include flowers such as bleeding hearts and violets and plants such as hostas and astilbes. Many perennials need to be divided as they mature, so any parents with perennials in their gardens might

be able to provide you with free divisions of the growing plants for your outdoor classroom.

Other Outdoor Classroom Components

Archaeological Dig

This is a sandbox in which teachers can bury "artifacts" and parts of skeletons for students to retrieve and assemble. The teaching objective is to show students how to use a grid to mark dig finds. Plastic skeletons are available from some science suppliers and often in toy departments or dollar stores as well. Clay flowerpots, broken into pieces that students can reassemble and glue, are also great items for retrieval.

Compost Bins

Choose different types of compost bins to use for lunch scraps and yard waste for comparison and contrast. Before you select bins, be sure to check your local ordinances. Many cities and towns have laws describing the various kinds of compost bins that are allowed within city limits. A simple bin can be made from inexpensive hog wire or chicken wire stretched around four posts that are secured in the ground. Compost bins can also be purchased from garden centers and science catalogs. If you have more than one bin, students can use nonmercury thermometers to compare temperatures in different bins.

Compass Rose

Use stones to make a permanent compass rose on the ground. A volunteer who is a brick mason might be willing to build a compass rose as a donation, perhaps involving your students in the process so that they could learn from his or her skills.

Human Sundial

Students can tell time by casting their own shadows on a series of markers on the ground. Use landscape timbers to mark each hour where a human shadow falls. You can verify placement with a watch and a compass. The person whose shadow you use should face north, and the student can increase the length of the shadow by holding both arms straight up above his or her head. This feature is a student favorite!

Pond

Life science classes benefit greatly from testing water samples, observing cells in water plants, and perhaps even viewing the life cycle of a frog in a pond. However, be aware that having a small pond requires an increased amount of maintenance. If you want a circulating pump to prevent water from becoming stagnant, you will also have the expense of electricity. A simple, less expensive solution is to build a shallow bog that is basically a muddy, damp place. You can use the bog to establish plants. (Safety note: Stagnant water attracts mosquitoes and other health hazards. Make sure students wear long-sleeve shirts and pants to reduce exposure. Ticks can also be a problem.)

Temporary Alternatives

What can you do if your plans have been turned down or you do not have the space, money, time, or energy to build an outdoor classroom learning lab? Fortunately, there are some practical, short-term alternatives. Here is how to adapt outdoor classroom components for temporary use.

Archaeological Dig

Look for a small plastic swimming pool on sale at the end of the summer. Fill it about halfway with sand and bury your "artifacts." A miniature archaeological dig can be set up in smaller plastic containers. Adjust the size of the "artifacts" to suit the size of the container.

Attracting Birds and Butterflies

Feeders for birds and butterflies can hang on shepherds hooks near your classroom window. You can also find feeders that attach to windows with suction cups. Butterflies are attracted to colorful feeders similar to hummingbird feeders. Fill these feeders with a mixture of four parts water and one part sugar. Ask parents to donate a birdhouse, which you should place in a tree in a convenient school yard location.

Compost Bins

Use plastic containers to compost food scraps (from plants) indoors. If you are on a strict budget, use small leftover plastic food containers—any plastic containers that can be sealed will do—and fill them with a variety of leaves and kitchen scraps. Be prepared for a strong odor when you open them. If you open the containers outdoors, the smell will be quickly diffused. In addition, some science companies sell worm farms or habitats, which would allow students to study how earthworms speed up the composting process. These products are made for indoor use. (Safety note: Make sure you know the source of compost material to prevent student exposure to sharps, etc.)

Container Gardens

You may think that nothing can beat a great outdoor garden, but container gardens can come in a close second, and their portability makes them acceptable to even the most conservative administrator. To make a lightweight container garden, fill the bottom of a deeper container with broken pieces of foam. Use high-quality potting soil for the best results, fertilizing the soil as needed. Existing flower beds on the school grounds are another option. With administrative permission, improve the look of these flower beds with nectar flowers to attract butterflies or some berry bushes to attract birds. Creating a container garden can also be a great way to start building a case for an outdoor classroom later, if that is your goal.

Math Patios

Find an open area in a parking lot, bus loading area, or walkway and have students measure a 10 ft. × 10 ft. square area. Mark it off in square feet with chalk to create an elementary hundreds chart patio. For middle school students who need a larger space, have students measure and mark a 20 ft. × 20 ft. square area. To make a math patio on grass, put stakes in the ground at four corners and create a grid with white string or bright yarn.

Safety Tips for Teaching in the School Yard

School safety guidelines are important inside and outdoors, but there are some additional precautions that can be taken to make the outdoor classroom or a lesson in the school yard safer for everyone. Here are a few tips based on my personal experience:

1. Communicate with the school office. Keeping a sign-up sheet for the outdoor classroom in the office will provide documentation of which teachers are outdoors. If the office has walkie-talkies for coaches who are outside, they should also loan one to you when you go to the outdoor classroom. If there are no walkie-talkies, take your cell phone and

have the front office number programmed in it. Make sure other teachers (next door to your classroom or across the hall) know that you are taking your class outside in case anyone is looking for you.

2. Be aware of student allergies to plants or insect stings. Read your students' permanent medical records and ask parents if their children have any allergies. As you communicate with parents, you may want to find out what action they take for insect stings at home, then make sure the school is prepared as well. Remember that a butterfly garden attracts other pollinators such as bees, so be aware and prepared.

3. Choose materials carefully. Working with concrete objects such as patio blocks, birdbaths, or benches puts heavy objects around little fingers and toes. (Safety note: Students should wear only close-toed shoes—no sandals or flip-flops.) This is not to say that these objects should not be part of the outdoor classroom; teachers should, however, consider potential injuries when choosing materials and deciding students' levels of involvement in building an outdoor classroom. Close adult supervision will prevent most accidents.

4. Play is for the playground. The outdoor classroom is a space for learning. This is something that all adults at the school must learn and enforce. Unstructured time in the outdoor classroom usually leads to students walking on top of benches, climbing on birdbaths, or behaving in other unsafe ways.

5. The outdoor classroom is not an outdoor eating area. Eating in the outdoor classroom can attract unwanted wildlife. Even

a spilled drink can attract insects that sting to seating areas. If you will have food in the outdoor classroom for any reason, make sure to pick up all trash and rinse the benches thoroughly with water.

6. Some words of caution are in order regarding animals mentioned in this book. Wild animals should not be kept as pets in classrooms or homes. Also, be aware that animals such as butterflies ordered from a science supplier should not be released into the wild in areas where they are not native.

7. Although more of a caution than a safety concern, teachers should consider carefully the choice of plants for outdoor classrooms. As a general rule, native plants are a good choice because they are ideally suited to the climate. Non-native plants that are carried by local nurseries can sometimes make good additions to the school garden. However, teachers should not introduce nuisance plants that would overpower local species. Get advice specific to your location from a botanist, extension agent, or master gardener.

8. Teachers should be aware of whether their state requires an Integrated Pest Management, or IPM, program so students will not be exposed to pesticides, herbicides, and other hazardous chemicals when working out-of-doors. Teachers should make every effort to ensure their school adheres to any state requirements and also be proactive in fostering an IPM program, even if the state does not require it.

9. An outdoor classroom should not be part of or even beside a playground. One is a place to play, and the other is a facility for teaching and learning. Teachers should make sure that the outdoor classroom is not located near a playscape and also that the selected area was not previously the site of a wooden playscape. On a related note, when working with "treated timber," teachers should make sure it was fixed with non-arsenic preservatives.

10. Make sure appropriate safety training and precautions are taken for all activities. For example, before doing activities that involve any hand tools, students should be trained in basic construction safety. Additionally, when working outside, students should use appropriate personal protective equipment (PPE), including safety glasses or goggles, gloves, close-toed shoes, hat, long-sleeve shirt, pants, sunglasses, and sun screen.

11. Students must use caution when working with sharp objects, such as clippers, scissors, clothes hangers, metal and wood stakes, and so on. Students should also be careful when working with timbers, which present possibilities of tripping or falling, sliver hazards, and so forth.

12. Students should wash hands with soap and water after activities, especially those involving seeds, soil, food scraps, yard scraps, and so on.

There are probably other safety considerations particular to your school yard or outdoor classroom. Keep a running list and discuss it with other teachers who take students outdoors. Revisit the list each school year to keep your students safe.

Does Money Grow on School Yard Trees?

Resources for Your Outdoor Classroom

You may find activities in this book that really excite you, but how will you pay for them? Money does not grow on trees—even trees in a school yard. The good news is that schools are nonprofit organizations that are eligible for thousands of dollars in grant funding each year. Schools can also benefit from many charitable individuals, corporations, community leaders, and parents.

Your challenge as an environmental educator is to channel grant funds, donations of goods and services, and volunteer help from parents and the community into projects to help your students participate in exciting outdoor lessons.

Although there is some degree of luck and chance involved in receiving donations and grant money, some tried-and-true ways can enhance your chances. An important skill for success is a persuasive writing and communication style. If you do not consider yourself a good writer, find a colleague whose writing skills you respect. Ask him or her to read over grant applications you have written or edit letters to parents or local businesses.

Another challenge that you must overcome is feeling discouraged when a request for donations is denied or an application is rejected. Grant-writing offers particular challenges to the teacher who wants to stay motivated and remain focused on success because of the rejection rate. Some grant writers say the rate of acceptance is frequently one in ten, but with practice you can increase your odds.

All of these efforts may seem like too much work in your already tight schedule. How can you motivate yourself to launch into fund-raising? Make it your goal to teach outdoor lessons that are fully funded by donations or grants. Tell yourself that teaching outdoors should not stress your own wallet. It should be a stress-free joy for both you and your students.

This chapter includes letters to parents, business partners, and community leaders that you can use or adapt to your needs as you seek money for your projects. The checklist on page 22 (Table 2.1)—along with the tips, list of resources, and advice in this chapter—will help you get your grant-writing effort organized and on the right track. With planning and thought, you can synchronize your fund-raising to bring together the resources you need for your outdoor teaching experience.

Parents can be your greatest ally in teaching outdoors if you keep them informed and involved. They can quash your plans if they do not understand what you are attempting to do. How can parents help?

- They can loan tools and donate supplies for the outdoor classroom building phase.

- They can provide plants from their own gardens and yards.

- They can donate their time to build, maintain, or add to an outdoor classroom.

- They can provide individual skills or resources from their jobs or companies.

- They can be cheerleaders for outdoor learning within the community.

Engaging parents' interest and help with your project will also help your students. Involved parents will be enthusiastic to hear news about outdoor classroom building and lessons, creating more enthusiasm among your students.

As with any letter home to parents, the letter about teaching outdoors should be both friendly and professional. The letter should help parents feel comfortable with the idea that their child will learn outdoors in unfamiliar ways. The letter should also make parents feel welcome to join you in some of the outdoor activities you have planned for your students.

Community leaders and business partners are also helpful sources of supplies and resources. How can members of the community help?

- They can donate money to buy expensive supplies or maintenance equipment.

- They can provide special services, such as installing benches or putting in patios.

- They can provide valuable networking connections that may lead to other involvement from the community.

- They can provide publicity (such as a local newspaper article) that may attract the attention of other potential donors.

Whenever you send out a letter, think carefully about the purpose of your communication with parents or business partners. Some sample letters are included at the end of this chapter. If a letter fits your situation, you can simply copy the letter, fill in the blanks, and send it out. These letters can also serve as models for more individualized letters that you write. The letters included announce that you are planning an outdoor classroom and ask for donations and volunteers. Choose the letter that best fits your situation, or combine them as needed.

Make sure your administrator first approves what you plan to do outdoors and the letters to parents or community leaders before they are sent out. It is best if your administrator is supportive and enthusiastic about stepping out of the box—or in this case stepping out of the building.

Grant-Writing

Does the idea of writing a grant proposal intimidate you? You are not alone. Let's be frank: Grant-writing can be frustrating. But it can also be rewarding—in fact, it can be one of the most positive experiences of your career. Why? A grant can help you provide experiences for your students that you never dreamed would be financially possible. This chapter provides you with the tools to write a successful grant. So dream big! Think about what you and your students can accomplish with some financial backing.

The tips in this book have come from direct experience with writing grants. One of the reasons grant-writing can be frustrating is because you get no specific feedback on grant applications, successful or not. You get either a "congratulations" or a "thank you for applying." If you really want to build on your successes, keep your own records. Compare a successful application to one that was not successful. What could you have changed to make the unsuccessful grant better?

Each time you apply for a grant, read two or three old applications that you have kept on file. That will jog your memory about the appeals and details that do or do not work. Some organizations that provide grant funding will give you summaries of successful

grants upon written request. Others provide a synopsis on a website. Take advantage of this opportunity and study the examples to see what you can do to polish your request and make your own application stronger.

If you can juggle the responsibilities of teaching school successfully, you can write a grant that gets funded. It may not happen the first time, but it will happen if you are persistent. As teachers, we insist that students do not give up, but we are apt to do so ourselves. Remember that somebody has to be the one person in ten who gets a grant. Decide today that you will be that person. Make it happen!

Grant-Writing Tips and Strategies

Before you begin searching for grant resources and writing applications, ask yourself some key questions and jot down your answers.

- What project am I planning?

- How will I accomplish the project?

- What materials do I need, and how much will they cost?

If you have answers for these three questions, then you have the backbone of a decent grant application. The goal is to take it from decent to dazzling so that those who read the application have no choice but to fund your request. Follow the Top 10 Tips on pages 20 and 21, and use the checklist on page 22 to help you accomplish this.

One way you can gain the confidence you need to write grants is to think about plans for your students that could require grant funding. Teachers who consistently plan "extra" activities for their students generally have what it takes to carry out those plans.

That enthusiasm is what will help you obtain the grant money you need, as long as your enthusiasm, knowledge, and need show on your application. Remember that the average acceptance of grant applications is only one in ten, so increase your odds by making the application excellent. You can also consider applying for more than one grant at a time. The worst that can happen is that the answer is no. Good luck!

Professional Growth Through Grant-Writing

Top 10 Tips for Grant-Writing

"Not failure, but low aim, is sin."
—Dr. Benjamin E. Mays

The following tips have been based on years of successful grant-writing and present a game plan that has consistently brought positive results.

1. Have a good plan.
Whatever you would like to do with the grant money you receive, make sure you have a well-planned project that either addresses a need at your school or enriches your students' experiences. The plan should address the "big picture" and cover all aspects of accomplishing your goal.

2. Match the need to the grant.
Whatever grant you apply for, make certain that it is intended for the type of project you have in mind. Some grants are more general than others, but most have guidelines that will let you know if your idea fits the grant's intentions. Be sure to do your homework; this will save time and prevent rejections.

3. Read the application carefully.
Read the application two or three times before you begin writing the grant. Familiarize yourself with the application by taking notes on separate paper as you read. What part of your idea goes where on the application? What words on the application give you clues about which features of your project you should highlight?

4. Get the approval of your administrator.
Before you spend every night for a week working on a lengthy grant application, make sure you share your idea with your principal. (Make a point of sharing the completed grant application later, too.) Most applications will require your principal's signature or a letter of recommendation.

5. Get started, and get started early.
Even if you have never written a successful grant before, write something to get it started. Your train of thought cannot develop if you do not get it in motion. One completed paragraph will encourage and excite you more than you think.

You should start on the grant application long before the deadline. Completing the application will take longer than you think, and the signatures and approvals after you finish writing will add time to the process.

6. Get personal advice from successful grant writers.
If you know any teachers who have received a grant, talk to them about it. Attend a grant-writing session at a professional meeting. There is no substitute for the one-on-one advice and exchange of ideas between experienced and novice grant writers.

7. Find a trustworthy colleague to offer proofreading and advice.

No matter how good a writer you are, you will need a great proofreader. You might need to go to a colleague who is an acquaintance rather than a friend to make sure you get true feedback from someone who can offer constructive criticism. Choose someone who has excellent communication skills. You need honest, professional, constructive criticism to make your application as strong as it can be. Once you get advice and suggested changes, be sure to make use of the proofreader when completing your finished draft.

8. Find a grant-writing buddy.

In addition to a proofreader, an experienced adviser, and a supportive administrator, look for a colleague who is at your level and work with him or her. Use e-mail, telephone, or face-to-face meetings to support each other through the grant-writing and application process. If you find a buddy from another school, you will be able to share information without competing against one another. Many grants are awarded to only one applicant per school during a school year, or sometimes you can only win once. If you and your buddy both have successful grants, trade copies of your applications. Then each of you can apply for the grant the other received during the following year.

9. Apply for two grants simultaneously.

You may not want to do this your first time around, but after you have had some success and experience with grant writing, choose a project that you want funded and apply for two grants at one time. You will have a better chance of getting the project funded, and you can follow your checklist and steps for both applications at the same time. Make sure to keep your papers separate, and you can double your chances.

10. Believe in yourself.

No matter how good a teacher you are, the grant-writing application can be hard on your ego and confidence level. Do not let it get to you. Many teachers who win grants have talent, but they also have unbridled self-confidence; that belief in themselves pulls them through the inevitable rejections on the road to their successes. Focus on your goals and believe that you have the staying power to see them through. If you do that, you have won half the battle.

Table 2.1. Grant-Writing Checklist

Write dates or notes in the blanks to keep track of your progress.

Task	Started	Completed	Signature	Other
Preliminary tasks				
Plan mapped out				
Grant matches need				
Carefully read application				
Administrative approval				
Practical deadline				
Proofreader lined up				
Getting it done				
Budget planned				
Writing the grant				
Matching funds				
Upload or mail application copies				
Results received				
File copy of grant				
Itemized receipts				
Secondary grant for more funds				
Publicity if grant funded				

Resources for Finding Grants

"It is not a disgrace not to reach the stars, but it is a disgrace to have no stars to reach for."
—Dr. Benjamin E. Mays

Grant money is available for nearly any project you plan for your school. You can search the internet, materials from professional organizations, information from professional seminars and conferences, educational journals, colleges and universities, and even some of the junk mail that you receive at school. The following list is by no means exhaustive, but nevertheless it is a good place to start for grants related to teaching outdoors. Many applications can be downloaded, and some can be completed online.

State and Local Resources

In every state and many local school systems, there are grant resources available only to teachers in that state or county. Look for links to state science organizations at the NSTA (National Science Teachers Association) website (*www.nsta.org*). Many of these groups offer state grants and teacher awards with cash prizes. For local resources, find the grants contact or coordinator in your school system. This person can usually give you ideas for resources. You might also contact your local United Way or county extension service to find out if these organizations offer grant funding.

Grants for Science, Math, and Gardening

www.kidsgardening.com

National Youth Garden Grants
Greenhouse Grants
Dutch Bulb Grants
Healthy Sprouts Award
($250 to $2,495 in garden products or cash grants)

www.nsta.org

National Science Teachers Association Teacher
        Awards and Competitions ($1,000 to $10,000 in
        cash grants directly to teachers)
Shell Science Teaching Award ($10,000)
Disney Planet Challenge
Grants from state chapters ($500 to $2,000)

www.nwf.org

National Wildlife Federation ($3,000 to $7,000)
Grants to nonprofits for "on the ground" efforts to save
        endangered species
Note: Partner with a nonprofit for this program.

www.paemst.org

National Science Foundation
Presidential Awards for Excellence in Science and
        Mathematics Teaching ($10,000)
Note: Cash awards to teachers who have demonstrated
        exceptional performance

www.toshiba.com/taf/

Toshiba American Foundation (Grants in varying
        amounts, averaging $5,000)

www.nctm.org

National Council of Teachers of Mathematics
        Mathematics Education Trust (MET) ($2,000 grants
        for classroom projects or continuing education)
Toyota TIME grants ($10,000)

General Grant Information

www.schoolfundingresources.org
www.schoolgrants.org
www.grantsalert.com
www.schoolfundingservices.org
www.ed.gov/free

Letter to Parents, Business Partners, and Other Donors
Donations Needed for Outdoor Lesson Supplies

Date: _____

Dear Parents,

During this school year, we will conduct a number of lessons outdoors to help our students use real-life situations as they learn about science, social studies, writing, and math. Research has shown that learning in the context of the natural environment raises student achievement. We would like to take advantage of the opportunities that await our students just outside the doors of our school.

To teach outdoors effectively, we need some items that are not in the school budget. When the cost of everything is added up, this seems like a costly venture, but most of these items are not expensive on their own. If every family could contribute one or two items, that would help us tremendously. Donated items can be new or used, but must be in safe operational condition.

The items we need are listed below.

_____   _____

_____   _____

_____   _____

_____   _____

_____   _____

_____   _____

If you have any questions about the items on the list or our outdoor lessons, please feel free to call me at _____. Please remember that donations are completely voluntary. Although the overall number of donations may affect how we carry out our plans, all students will take part in our outdoor lessons, regardless of whether their families make donations.

Thank you for considering a contribution. I hope that you will hear from your child about how we used your donations or have the chance to see us firsthand as we work in our outdoor classroom.

Thank you,

_____

Signature

Letter to Parents
Outdoor Volunteers Needed for Work Day

Date: _____

Dear Parents,

As you know, we are conducting some of our lessons outdoors this year. Research has shown that learning in the natural environment raises student achievement, and we want your child to have this advantage. To make this possible, we need your help. Parents can help us by volunteering to complete the tasks listed below. Parents and students will have the opportunity to work side by side. If you have tools and work gloves that you can bring, that will be helpful. Please note that all tools and other equipment must be in safe operational condition.

(Return bottom portion.)

- -

Tasks to complete: (Check if you can help.)

☐ _____

☐ _____

☐ _____

Tools we need to borrow: (Check those you can bring.)

☐ _____   ☐ _____

☐ _____   ☐ _____

☐ _____   ☐ _____

Name of Parent _____

Student _____

Daytime Phone _____

Home Phone _____

If you do not have tools that we can borrow, we can still use your time and energy! Please fill out the bottom half of this letter to let us know how you can help. Feel free to call me at _____ if you have questions.

Thank you,

_____

Signature

Letter to Parents—Building an Outdoor Classroom

Date: _____

Dear Parents,

Research has shown improved student achievement when the natural environment is used as a context for learning. Our school is planning an outdoor classroom in which students can study in garden areas designed to fit our curriculum.

As you can imagine, an outdoor classroom can be costly, but we hope that with the involvement of parents and the community the project can be done without additional financial burden on the school. You can help us by giving your time, services, extra plants from your yard, or gardening tools.

If you have plants that you would like to share, please list them below so we can fit them into our garden plan. If you would like to be on the outdoor classroom volunteer list, please sign up to help in that capacity. If you are handy with woodwork, perhaps you can build some birdhouses or bird feeders for us. If you have experience in gardening or another environmental area, you could advise us as we work on our outdoor classroom or provide us with information for our lessons.

Please let us know below how you can help and return the bottom portion of this letter to our classroom. Feel free to call me at _____ if you have questions. Any donation you can make to our outdoor classroom is greatly appreciated!

Sincerely,

_____

Signature

(Return bottom portion.)

- -

☐ I would like to be on a list of outdoor classroom volunteers.

☐ I can donate _____ _____ plants.
              (Number)   (Type)

☐ I can provide these services: _____

Name of Parent _____ **Student** _____

Daytime Phone _____ **Home phone** _____

Letter to Business Partners and Community Members
Assistance Needed for Outdoor Lessons

School _____   **Phone** _____

Date: _____

Our school is taking on the task of teaching environmental lessons outdoors in the school yard. We plan to create an outdoor learning experience for students to enhance science, math, and language arts lessons.

This is an expensive proposition, but we know that with the help of community leaders and business partners we can accomplish this goal. That is why we are coming to you for support of these important educational goals.

As we have reviewed our plans for teaching outdoors, we thought that you may be able to help us with the following goods or services:

_____

_____

_____

Donated items can be new or used, but must be in safe operational condition. Any donations you make to our school, a nonprofit organization, are tax-deductible. We hope that you will look favorably on our request. If you are unable to give everything that we need, perhaps you can still fulfill part of our request. We also would welcome volunteer hours from your employees. We will contact you personally to follow up, but please feel free to call us at _____ if you have questions.

Thank you for your support of our school. Together, schools and business partners can build a better workforce for the future.

Sincerely,

_____

Teacher

_____

Principal

Birds, Bugs, and Butterflies

Science Lessons for Your Outdoor Classroom

Outdoor classes give teachers the chance to share nature's wonders with students firsthand. From the smallest insect to the largest flying predator, children will retain far more knowledge if they watch animals in natural settings instead of simply reading about them.

One of the most important concepts to share with students is that animals have four basic needs: sources for food, a water source, shelter, and space to bring up their young. When planning an outdoor learning experience, think about how many of these resources are already available on the school grounds and what you can add to make the environment more hospitable for wildlife.

Among the wild animals that may travel through a school yard, birds, bugs, and butterflies are the most common—the focus of most of the lessons in this chapter. There is an abundance of information for teaching about these animals available through the internet, books, and nature centers.

As you develop your outdoor-teaching skills, you may want to bring a wider variety of wildlife into your lessons. You must take into consideration the safety of your students and the location of your school when thinking about attracting wildlife. Animals that move through the air are less of a nuisance and therefore more acceptable in the

school yard and surrounding area, regardless if the school is located in an urban or rural setting.

For many birds, shelter could be as simple as trees that already live in your school yard. Water might be available in a stream or ditch, or you could easily set up a birdbath. Depending on the species, a food source might be insects living naturally in the area, seeds from flowering plants, or several bird feeders that can be viewed from your classroom window.

Butterflies need flowers for their food sources as adults and host plants as a food source for caterpillars (and a place to raise their young). They also need shallow pans of sand and water for "puddling" and shrubs, fence posts, or trees on which they may climb to form their chrysalides.

Finally, space or a place to raise young must not be overlooked. The more space you have, the more area is available for animals to estab-

lish their own territories, which is particularly important so that students can compare the habits of various species. You can increase the school yard's appeal to certain birds by adding birdhouses. Many students will enjoy building and placing the houses as well.

The following activities may lead to establishing permanent additions to your school yard or having your school yard certified by the National Wildlife Federation as a Certified Wildlife Habitat (*www.nwf.org*). Even if this does not become a permanent addition, however, sharing these close-up views of animal environments will teach students valuable lessons about being informed citizens in the natural world and make science concepts more meaningful, fun, and easy to recall.

This chapter offers a variety of activities to allow you to "tame" the wildlife to help you teach. Instructions for each lesson are presented first to help you make the most of each handout. Handouts for the activities start on page 33.

Don't Forget!

- Communicate with the school office. Make sure other teachers (next door to your classroom or across the hall) know that you are taking your class outside in case anyone is looking for you.

- Be aware of student allergies to plants or insect stings.

- Choose materials carefully. Working with concrete objects such as patio blocks, birdbaths, or benches puts heavy objects around little fingers and toes.

- Play is for the playground. The outdoor classroom is a space for learning.

For a full list of safety tips, see pages 15 through 16.

Animal Habitat Survey

Teaching Objectives: to identify the basic needs of food, water, shelter, and space required by animals to raise their young; to identify resources to meet the basic needs of animals living on the school grounds

Why/How to Use This Lesson: In any study of populations and ecosystems (NRC 1996, p. 140), an activity such as this study of the local environment helps students form a local perspective on a global concept. Use this lesson to help students sharpen their observation skills and gain a greater understanding of how local organisms fit into their ecosystems.

Materials: handout, clipboard, pencil

Procedures and Tips: For this lesson, it would be helpful for you to take a walk through the school yard before taking your students outside. Look for specific examples that meet the needs of animals. Make notes about possible animal habitats.

Before leading the class outdoors, engage students in a discussion of animals' basic needs. Animals need food, water, shelter, and space

to raise young. Discuss these needs and relate them to the resources on the school grounds.

First, talk about what animals eat and drink. Some examples to mention might be that squirrels eat nuts and seeds from trees, spiders eat insects, and caterpillars eat leaves. Water sources for animals can include mud puddles, water dripping from a gutter on the building, streams, or birdbaths.

Shelter for animals can be trees, the eaves of the school building, and rocks or underbrush. Space for raising young connects to shelter and can include trees for some animals and open fields for others.

Assessment/Next Steps: After students have had a chance to explore outdoors and identify a number of resources that animals need, gather students back into a group to discuss the results of their school yard survey. If you are still planning your outdoor classroom, this activity will allow students to help you adjust and improve the plans. What have they found that meets animals' basic needs, and what can they identify that is lacking? If they have answered this question adequately, then they have mastered the intended concept.

SCiLINKS.
THE WORLD'S A CLICK AWAY

Topic: What Is a Habitat?

Go to: www.scilinks.org

Code: OS001

Animal Habitat Survey
Basic Needs of Animals—Resources in Our School Yard

Name: _____ Date: _____

Directions
Take a walk on the school grounds with your teacher. Complete this form to show what resources are available to animals.

1. What sights and sounds do you see or hear that tell you animals are here?
   _____
   _____
   _____

2. What proof can you find that animals have been here? (e.g., footprints, anthills, droppings, feathers, etc.)
   _____
   _____
   _____

3. What animals may have lived here before there was a school building? What was taken away from the environment that may have made them leave?
   _____
   _____
   _____

4. Compare your findings with a partner's findings and talk about the basic needs of the animals you wish to attract. Then check your findings with those of the rest of the class before filling in the chart below.

Basic Needs	Resources in Our School Yard

How Birds React to Environmental Changes

Teaching Objective: to make predictions and inferences regarding the responses of birds (and other animals) to changes in their environment

Why/How to Use This Lesson: Use this lesson to support units of study on the traits and behaviors of organisms (NRC 1996, p. 129). You can also use this lesson to support the fact that individual animals do react to environmental changes but adaptation and natural selection occur over generations. This lesson should help students define adaptation as "variations organisms are born with that can lead to an individual's survival and reproduction" (Keeley, Eberle, and Tugel 2007, p. 144).

Materials: handout, clipboard, pencil, binoculars (optional, for bird watching), "Habitat Change" assessment probe (Keeley, Eberle, and Tugel 2007, p. 143)

Procedures and Tips: Discuss animals and how they react to changes in their environments before you take students outdoors. You may wish to read out loud from a resource book or show a video from the internet to offer students background information. Consider using the formative assessment probe listed in the Materials section above.

When going outdoors to observe birds, remind students to be quiet and still. Birds will frequently be scared away by noise and movement. One way to view birds with an active class is to sit far away from the spot where birds gather. You will need binoculars for conducting the activity this way. If you do not have bird feeders, search the school yard for trees where birds live. You may be able to spot nests ahead of time and find the best place on school grounds to take your students.

If you have little luck with natural attraction of birds to the school grounds, try scattering mixed birdseed in a field or along the edge of a wooded area. If you see that this works, make it part of the observation by letting your students in on the strategy. Compare what happens when the seed is scattered to when it is not.

Assessment/Next Steps: Evaluate students' answers to questions on the handouts to assess understanding. Consider taking this lesson to the next level by introducing adaptation and natural selection.

How Birds React to Environmental Changes

Name: _____ **Date:** _____

Directions

Animals have to make adjustments to survive changes in their environments. Birds often have to react to changes such as new buildings, trees being chopped down, or less food in the area. Think about birds that live in your area. Watch the birds that live on the school grounds and then fill in the blanks below.

1. Write down three changes that these birds have made to survive.

   a. _____ b. _____ c. _____

2. Compare your list with a partner. Add something from his or her list.
   If your lists are the same, brainstorm one more change.

   _____

3. Next, think about animals' four basic needs. Complete the chart below to show birds' needs and reactions to their surroundings. A few examples are provided.

Think About It!

Is migration an option when the habitat changes? Why or why not? Explain your answer on the back of this page.

Basic Need	Source That Meets Need	Reaction When Need Is Not Met
Food	(e.g., hunts for seeds in fields or woods)	(e.g., gets seeds from bird feeders)
Water		
Shelter	(e.g., builds nests in trees)	(e.g., trees are cut down; lives in birdhouse)
Space		

The Great American Backyard Bird Count

Teaching Objectives: to engage students in a wildlife counting activity; to connect students with the scientific community

Why/How to Use This Lesson: In a unit on regulation and behavior of organisms (NRC 1996), it may be helpful for students to gather information in much the same way that scientists do. This activity would fit into units on migration, animal habitats, food webs, or food chains.

Materials: handout, clipboard, pencil, binoculars, internet access (visit *www.birds.cornell.edu*)

Procedures and Tips: If your school has a computer lab, have your class visit the website for Cornell's ornithology lab (see Materials section). Even if you do not have internet access for your students, visit the Cornell University's website yourself for additional background information before you teach this lesson.

Counting birds and other animals is an activity that scientists have conducted throughout modern times and students will find simple and engaging. This activity will also benefit from your scoping out the bird populations in your school yard ahead of time. A preview will give you an idea of some reasonable numbers to expect from your students when they do their counts.

If possible, place a number of bird feeders on your school grounds to increase your chances of seeing a variety of birds. Consider offering sunflower seeds in one feeder, thistle seed in another, and mixed seed in a third feeder. You may even spread seed on the ground if feeders are not in your budget. Actually, certain kinds of songbirds are ground feeders. As with the activity "How Birds React to Environmental Changes," involve the students in attracting the birds to the school yard. Let students formulate a plan for what kind of seed to spread based on research in field guides and on the internet.

You can make a temporary bird feeder by rolling a pinecone first in peanut butter and then in birdseed. Tie it to a tree branch with string.

Assessment/Next Steps: After your students have conducted their observations and compared them in small groups, facilitate a class discussion about their discoveries. Ask students if the results might be different in another season. (Migration would affect fall and spring numbers in many areas.) If you plan to conduct the activity again, save the handouts and compare students' results the next time. This will yield an opportunity to graph results if students observe differences.

The Great American Backyard Bird Count

Name: _____ Date: _____

Scientists at universities all over the world study animals. Some of the scientists at Cornell University study ornithology, the science of birds. (Visit *www.birds. cornell.edu* to find out more.) You can participate in the bird-counting programs that Cornell University offers on its website or conduct a count of your own.

Directions

Take a trip outside or watch birds from inside the classroom. Look for different kinds of birds. Then answer the questions below.

Tip: It is helpful to sit near bird feeders for this activity. Check to see if the feeders are filled with various kinds of seeds. If you do not have bird feeders on your school grounds, try sitting near some trees. Be still and quiet so that you do not scare the birds.

1. In a 10-minute period, how many birds did you see? _____ birds

2. How many different species of birds did you see? (You do not have to know the names to answer this question. Just observe that they are different types.)
   _____

3. Which bird species did you see most often? (If you do not know the name, describe it now and check in a field guide later.) _____
   _____
   _____

   How many of this species did you see?_____

   Form groups of three students and compare your answers.

   Who saw the greatest number of birds? _____

   How many? _____

4. Would adding your numbers together be an effective way to get the total number of birds in the school yard? Why or why not?
   _____
   _____
   _____
   _____
   _____
   _____
   _____
   _____
   _____

Do You Hear What I Hear?

Teaching Objectives: to identify the role of animal sounds in nature; to compare and contrast the sounds made by various animals

Why/How to Use This Lesson: A lesson on animal sounds fits into the study of animal regulation and behavior (NRC 1996), particularly behavior as it relates to environmental stimuli and communication. Animals may use sounds to warn of predators or find a mate. When planning a lesson on animal sounds, consider ways to develop the students' understandings of how the sounds help animals survive.

SCI LINKS.
THE WORLD'S A CLICK AWAY

Topic: Animal Communication

Go to: www.scilinks.org

Code: OS003

Materials: CD or tape of nature sounds, CD player or stereo, handout, clipboard, timer or watch with a second hand

Procedures and Tips: Shop around for CDs or recordings of animal sounds that you can play for your class. Some of these recordings have themes such as rain forest or ocean. Listen to the recording ahead of time and choose the part that has the most distinct animal sounds. If possible, use a recording that has sounds native to your area.

To set the mood when you play the sounds for the class, turn the lights low and have students close their eyes. Ask students to listen for at least two minutes (more if possible). Ask them to write down as many different animals as they can identify. Ask them to be specific about animals (for example, students should not say "birds," but instead specify owls, seagulls, hawks, and so on).

As in other lessons, preview the school yard to determine if there are sounds that are audible and plentiful. (If there are no audible sounds, try listening outdoors at home at night, and consider making this a homework assignment.) After students have listened to recorded and real animal sounds, have them discuss why animals make sounds. Research the topic with books from your media center or on the internet. If you have access to a recording device (such as a cell phone, laptop computer, or graphing calculator sound probe), you can try recording nature sounds at night to share with the class.

Assessment/Next Steps: Assess students' understanding by evaluating the paragraphs they have written from the Reflect section on the student handout. After this lesson there are numerous possible extensions, such as comparing the sounds of nature in different biomes, at different times of day, or in various places in your community.

Do You Hear What I Hear?

Name: _____ **Date:** _____

Evidence of animals in the wild comes in many forms. Have you ever thought of sounds as evidence? Even though you cannot hold the sounds in your hands, you can record them electronically. You can also keep a written record of what you hear. Go outdoors with a partner. One person will close his or her eyes while the other person takes notes. The student who has his or her eyes closed will listen first for bird sounds and then for frog sounds. The writing partner will write what sounds the partner hears. Work together for 15 minutes, taking turns. Fill in the chart below. Be sure to talk quietly so you do not scare your test subjects!

Directions

Use a clock, timer, or stopwatch to time yourself. Every time you hear a bird or frog sound, tell your partner. The partner who is writing should make a mark in the correct column below. If you do not have birds or frogs on your school grounds, your teacher may let you choose other animal sounds to observe.

Time Interval	Bird Count	Other Animal	Frog Count	Total Sounds

Reflect

Think about the different sounds you heard on the recording your teacher played for you. Think about the sounds you heard when you listened outside. Think about the other animals you have heard, too. Why do animals make sounds? Do they respond to one another? Imagine animal sounds from another biome, such as the rain forest or ocean. What would they sound like? Write a paragraph to answer these questions on the back of this page.

Internet Connection

Find animal sounds on the internet. For frog sounds, search for the species spring peepers. See if you can find the sounds of at least three different animals to share with your class.

3

The Migration Sensation

Teaching Objectives: to identify animals that migrate through the local area; to provide an understanding of the advantages of migration

Why/How to Use This Lesson: In a unit on animal behavior or perhaps an integrated unit on migration, this lesson puts into perspective the advantages and disadvantages of migration. If you introduce natural selection, students can begin to realize that animals with the instinct and ability to migrate have survived over time.

SCiLINKS
THE WORLD'S A CLICK AWAY

Topic: Migration
Go to: www.scilinks.org
Code: OS004

Materials: handout, pencil, clipboard, resources for research, paper for brochure, bird feeders for migrating birds (optional), flowers for migrating butterflies (optional), digital camera for brochure pictures (optional)

Procedures and Tips: If your students have completed the activity "How Birds React to Environmental Changes," then they will have a head start on this lesson because migration can be taught from the perspective of being the ultimate response to environmental change. In other words, migrating animals move to a new environment when the current one becomes difficult. However, it is important to note that migration has occurred across generations. Entire species migrate—not just one bird who "decides" it is too cold and therefore flies south for the winter. After discussing this concept with the class, take a walk through your school yard or outdoor classroom. Have your students take their handouts on clipboards to answer questions 1 and 3. Question 2 can be done after you go inside. As you walk through the school yard, tell students to think about how they might make a travel brochure for migrating animals. If you have access to a digital camera, students may take pictures to use in their brochures.

Discuss the answers in cooperative groups or as a whole class, then have students work on their brochures. Some computer software programs have templates for brochures, which may help students with design ideas.

Assessment/Next Steps: To assess understanding, evaluate students' brochures to determine if they include several advantages of migration. To take this further, use the internet (*www.learner.org/jnorth*) to find target animal species that students can study in-depth.

The Migration Sensation

Name: _____ **Date:** _____

Migration is the seasonal movement of animals to a completely different location. They migrate to survive and meet their basic needs. Among the animals that migrate are salmon, whales, monarch butterflies, and many different kinds of birds.

Directions
Answer the questions.

Migration Through the School Grounds

1. Find out what animals migrate through your area. Spend 10 to 15 minutes making observations. Write down what you see and hear. What animals do you think might migrate through your school yard? What makes you think so? (Hint: Look for evidence! Do you see some birds only at certain times of the year? Do you see animals gathering to migrate? Are there animals in your school yard that might not want to spend the winter in your area?)

   _____
   _____
   _____
   _____

2. Back up your theory above with facts. Research migration in your media center. Look up the Journey North website (*www.learner.org/jnorth*). Check to see if you were right. What are some animals that actually migrate through your area?

   _____
   _____
   _____
   _____

3. How do the school grounds meet the needs of migrating animals? What can you and your classmates do to make the area a better place for migrating animals?

   _____
   _____
   _____
   _____

Show What You Know
Pretend you are a travel agent. Make a brochure that shows your school yard as a travel destination for migrating animals. What can you offer these animals? Read the travel section of a newspaper, a travel site on the internet, or brochures from a local travel agent for ideas and phrases. Use these to talk about your school yard's best travel features. Be creative! Attract nature's "tourists" to your school!

Monarch Butterfly Life Cycle

Teaching Objective: to identify by sight the stages of the monarch butterfly and its host plant

Why/How to Use This Lesson: This activity would be a good first lesson for a monarch unit, whether you decide to raise monarch caterpillars in your classroom or search for the adult butterflies and their offspring on milkweed plants. It could also fit into a general unit on life cycles.

Materials: handout; crayons, colored pencils, or water-based markers; internet access (visit *www.monarchwatch.org*); "Does It Have a Life Cycle?" assessment probe (Keeley, Eberle, and Dorsey 2008, p. 111)

Procedures and Tips: Consider using the assessment probe listed in the Materials section to start a unit or discussion about life cycles.

There are about two dozen species of milkweed, including common milkweed, tropical, swamp, sand vine, and narrow leaf. A complete guide to milkweed, including photographs and information about various species, can be found on the Monarch Watch website (*www.monarchwatch.org/milkweed/guide/index.htm*).

More than just a coloring exercise, the handout can be used to show the monarch and the host plant as they appear in nature. Use the multimedia gallery on the Monarch Watch website to make sure all of the stages depicted on the handout are colored correctly. There are various shades of orange in the monarch wings, and it may be a good time to mix in a lesson about creating different shades of paint colors. Provide students with crayons, water-based markers, and colored pencils instead of limiting them to one medium.

Explain to the students that the adult butterfly drinks the nectar from the flowers and lays eggs on the underside of the milkweed leaves—only one egg per leaf. The caterpillar eats the leaves. Emphasize that milkweed plants provide food, shelter, and a place to raise monarch young.

Assessment/Next Steps: To assess learning, answer this question: Did students correctly label the stages and closely match the colors from photographs of the monarch life cycle? To take it further, have students compare and contrast the stages with those of other butterflies or living things.

Monarch Butterfly Life Cycle

Name: _____ **Date:** _____

Directions
Fill in the blanks to label the stages of a monarch butterfly's life.

Word Bank
adult butterfly

caterpillar

chrysalis

egg

hanging "J"

milkweed

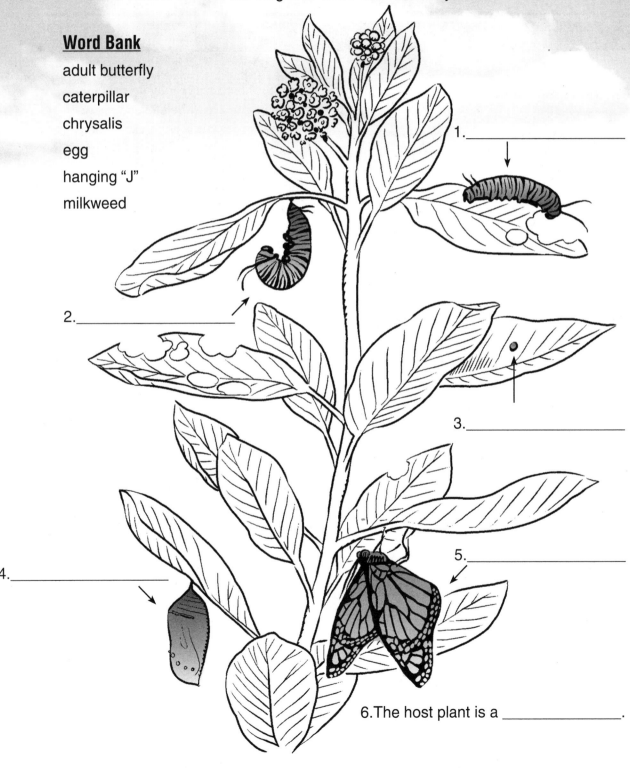

1. _____

2. _____

3. _____

4. _____

5. _____

6. The host plant is a _____.

What Do Swallowtail Caterpillars Swallow?

Teaching Objectives: to identify host plants of caterpillars; to make observations and inferences about caterpillars' preferences for certain plants

Why/How to Use This Lesson: This lesson fits into the study of animal habitats. When students help teachers plan animal habitats for the school yard, a comparison of which plants attract more animals can help determine future plantings. This is another opportunity for students to gather and analyze data, which are everyday activities for scientists in the field.

Materials: fennel, dill, and parsley plants; swallowtail caterpillars; handout; clipboard; pencil

Procedures and Tips: Swallowtail butterflies are one of the most common in North America. There are many varieties. Most prefer one of three herbs as host plants: fennel, dill, or parsley. With these plants, you can test the preferences of swallowtails in your area. It may even help you determine the variety of swallowtail that is native to your state.

Have students complete the following steps:

- Germinate seeds to establish plants in the garden area or purchase plants.

- Put plants in a sunny area along with some flowers for the adult butterflies. (To do this activity indoors, purchase swallowtail chrysalides from your favorite science vendor and put the mature plants inside an insect cage.)

- Observe the area until swallowtails are spotted. If none are seen, check the leaves of the plants for small white specks, which are the butterfly eggs.

- When the eggs hatch, keep a record of which plant is eaten. If more than one plant type displays eggs, compare the amount and the rate at which caterpillars eat the leaves of each plant after they hatch.

- Use the chart on page 45 to record the results. Analyze the data on the chart to determine what the swallowtails in your area prefer.

If your experiment is not successful (no butterflies were attracted and no eggs laid), research additional species of butterflies native to your area. Check a butterfly field guide, or use the internet (*www.butterflywebsite.com*) for ideas.

Assessment/Next Steps: Evaluate the student answers to questions below the data chart on page 45. Did students support answers with valid reasoning? To take this further, compare the eating habits of various species of caterpillars or follow up with the lesson on adult swallowtail butterflies (What Do Adult Swallowtail Butterflies Swallow?).

What Do Swallowtail Caterpillars Swallow?

Name: _____   **Date:** _____

Directions
Look carefully at swallowtail host plants. Record your observations on the chart below.

Plant	Date planted or date seeds sprout	Date eggs are laid / number of eggs	Date eggs hatched & number of larvae	Estimate of percentage of plants eaten in one week
Fennel		/		
Parsley		/		
Dill		/		

Note: One way that scientists measure the amount of food that a caterpillar (larva) has eaten is by collecting and weighing frass, or caterpillar waste. You may not want to do this, but you can look for green pellets under the plant. If you see frass underneath it, then that is what the caterpillar has been eating.

What can you learn from the data in the chart?

1. The caterpillars preferred to eat _____ because _____
   _____
   _____
   _____

2. If the caterpillars preferred one plant over the others, what does this mean about the other two plants? Support your answer. _____
   _____
   _____
   _____

3. Fennel and dill are more likely to flower than parsley, but did you notice any butterflies drinking nectar from any of these plants? _____
   Why do you think they did or did not? _____
   _____
   _____

3

What Do Adult Swallowtail Butterflies Swallow?

Teaching Objectives: to make observations and inferences about animal feeding habits using butterflies as an example; to identify the feeding preferences of certain butterflies

Why/How to Use This Lesson: Teachers may use this lesson as part of a unit on structure and function of living things. The butterfly's proboscis is unique and often sparks student interest in how various animals eat. This lesson pairs nicely with another lesson on how birds use their beaks to pick up various types of food. Teachers may consider including this lesson idea as part of the study of food webs and food chains because the plant is a producer and the butterfly is a consumer.

Materials: three varieties of flowering plants, butterflies, handout, clipboard, pencil

Procedures and Tips: Use a container, school yard flower garden, or space in an outdoor classroom to plant three or more types of flowers. Then have your students observe the flowers to determine which ones are visited by butterflies most frequently.

Prepare students for this activity by watching a video of a butterfly using its proboscis (feeding tube) to drink from a flower. Students will then know what to look for while making their observations in the garden.

This activity will work best if done over a period of time, with students gathering information at various times over several days or weeks. Weather and luck will play a part in this activity, so you will have to be flexible enough to make time for observations on sunny days when butterflies are active in the school yard. If you would rather teach this lesson indoors, a butterfly habitat with three types of flowers inside will serve the same purpose and perhaps offer simpler observation and more predictable results.

Assessment/Next Steps: To evaluate understanding, read students' responses in the Butterfly Visits section above the data chart. Did students support their reasoning for their observations? To take it further, have students illustrate or model a food web or food chain, including the swallowtail and flowers in the study they conducted.

What Do Adult Swallowtail Butterflies Swallow?

Name: _____ **Date:** _____

Directions
Adult butterflies drink flower nectar for their food. See if there is one type of flower that the butterflies like better than others. Observe swallowtail butterflies and complete the chart.

- Record the types of flowers in the first column. Make a mark for each time a butterfly lands on the flower.
- If you see the butterfly's proboscis unfurl, highlight the tally mark using a water-based marker.

Butterfly Visits
Put yourself in the butterfly's place. Write in the first person (or first insect, to be exact!) to explain the reasons for why you made the choices. (*Example:* I preferred the nectar of the purple coneflower because . . .) Continue on the back of this page.

_____
_____
_____
_____
_____
_____
_____
_____
_____
_____
_____
_____
_____
_____
_____
_____
_____
_____
_____

Flower	Day 1	Day 2	Day 3	Day 4	Day 5
A					
B					
C					
Example: Marigold	I I I	I I	I I I I I I I	I I I I I I I I I	I I I I I I

3

Milk a Weed for All It's Worth!

Teaching Objectives: to identify various types of host plants for monarch butterflies; to gather and graph quantifiable data from observations in an experiment

Why/How to Use This Lesson: This activity may help students understand how plants and animals interact. It would be helpful in a unit on food webs or an integrated unit on monarch butterflies.

Materials: three varieties of milkweed seeds or plants, monarch butterflies or caterpillars, flowerpots, potting soil, water, trowel, handout, pencil (Safety note: Potting soil may contain venmiculite, which has asbestos fibers. Keeping potting soil moist significantly reduces exposure to asbestos fibers.)

Procedures and Tips: Order three varieties of milkweed seeds from Monarch Watch (*www.monarchwatch.org*). Germinate the seed indoors and transplant the plants outdoors when they are about 20 cm tall. Have students create a chart to gather data. Make sure the data are quantifiable (able to be translated into numbers for comparison). Students may choose one variety of milkweed as a comparison and keep it indoors in a cage with a butterfly raised from larva to adult. You may wish to conduct the entire activity indoors by placing pots of all three milkweed types in the insect cage with monarch butterflies.

Assessment/Next Steps: Evaluate students based on their interpretations of the quantitative data they have collected. Can students explain their reasoning about conclusions they made using the data? For additional studies, have students research what other insects call milkweed home. There are additional food chains involving aphids, ladybugs, and ants.

Milk a Weed for All It's Worth!

Name: _____ Date: _____

On most milkweed plants, adult monarch butterflies drink the nectar of the flower. Meanwhile, the caterpillars eat only the leaves on a host plant. This makes for "one-stop shopping" when the monarch is on the go!

Directions

Compare the varieties of milkweed available in your area to find out if monarch butterflies have one they like better than others. Locate three different kinds of milkweed plants if possible. Create a pictograph to track how many times the monarchs visit the plants. Draw a butterfly shape on your graph for every five times the monarchs drink nectar from each kind of milkweed. Observe the milkweed plants over several days.

Note: The plant that is most eaten is the favorite variety of the caterpillars.

Draw one butterfly shape to represent 5 visits to the plant.

Plant A	
Plant B	
Plant C	

Results

Which kind of milkweed should you plant in the school garden? Why?

_____
_____
_____
_____
_____
_____
_____
_____
_____
_____
_____
_____
_____

3

Raising My Caterpillar

Teaching Objectives: to practice making predictions based on evidence; to observe the life cycle of a butterfly; to care for a living being

Why/How to Use This Lesson: Raising a caterpillar helps students develop conceptual understanding of life cycles by giving them a concrete example to follow. It is a good way to start a unit on insects, life cycles, butterflies, or even the needs of living things.

Materials: caterpillars, small plastic container, insect cage, food source (the correct host plant leaves) for caterpillars, paper towel, stick, ruler, handout, pencil

Procedures and Tips: The easiest way to get started is to order a butterfly life cycle kit from a reputable science supply company. One of the most common butterflies sold as larvae is the painted lady. These caterpillars are hearty, and students can care for them easily. They come with their food medium, containers, and directions. However, if you find caterpillars that you can identify reliably by species and correct food source, you can raise them without a kit. Be sure to provide students with small containers. A small plastic peanut butter jar is just about the right size. Punch holes in the top and cut a piece of paper towel to put between the lid and the lip of the jar so the caterpillar will not hurt itself trying to get out of the holes. The paper towel also makes a convenient place for the caterpillar to make its chrysalis.

Have students clean the waste out of the bottom of the jars each day and replace uneaten food with fresh leaves. (Safety note: Try to keep the insides of the jars dry to prevent mold growth.) Each student should observe his or her caterpillar each day and record the caterpillar's size. Keeping a science journal to describe the changes is an important step. Students can measure caterpillars through the sides of the jars most easily when the caterpillars walk up the inside of the jars.

When each caterpillar makes its chrysalis on the paper, take the paper out and wrap it around a stick. Place the stick in an insect cage. When the butterfly emerges, usually a week or so later, observe it for one day and then release it outdoors. (*Note:* Make sure when you order butterfly larvae from a science supplier that the species is native to your area.) After all the native butterflies are released, have students compile their data and represent it on graphs.

Assessment/Next Steps: The data should represent reasonable findings and be recorded accurately in graphs. To extend the experiences in this lesson, you could provide opportunities to compare the caterpillar and butterfly life cycles with those of other animals that go through metamorphosis, such as frogs or mealworms and beetles. You could also have students research particular caterpillars. There is a great deal of information available on the monarch caterpillar, for example, on the Monarch Watch website (*www.monarchwatch.org*).

Raising My Caterpillar

Name: _____ **Date:** _____

Directions

Take care of a caterpillar that is given to you. Caterpillars only need a small container. There should be tiny pinholes in the lid for air, and you must keep the container clean. Clean the caterpillar's home every day. Put in fresh host-plant leaves each day. Complete the sentences below.

1. My caterpillar is the larval stage of a _____ butterfly.
   (species)

2. I got my caterpillar on _____ when it was _____ centimeters long.
   (date)

3. The host plant for this type of caterpillar is _____ .

4. The color of the caterpillar on the first day is _____ .

5. Make a drawing to show how the caterpillar looked on the first day.

6. I predict that the caterpillar will make its chrysalis on _____ .
   (date)

7. After three days, the length of my caterpillar is _____ centimeters.

8. I have made these three observations of my caterpillar in the first three days.

   a. _____

   b. _____

   c. _____

9. I predict my caterpillar will be _____ centimeters long in three more days.

Show What You Know

After measuring the growth of your caterpillar every day, make a graph of the growth on the back of this page. Remember to label all parts of the graph.

Animals Living on the School Grounds

Teaching Objective: to compare and contrast animal adaptations in the local environment

Why/How to Use This Lesson: This activity is the culmination of those earlier in this chapter. Conduct at least three or four other school yard activities first (such as Animal Habitat Survey, How Birds React to Environmental Changes, The Great American Backyard Bird Count, and Do You Hear What I Hear?).

SCI LINKS.
THE WORLD'S A CLICK AWAY
Topic: Adaptations of Animals
Go to: www.scilinks.org
Code: OS007

Materials: handout, clipboard, pencil, digital camera for photographs of animals in school yard (optional), binoculars (optional)

Procedures and Tips: Students should use data from other activities: charts, handouts, graphs, or any other data from a science journal or notebook. Then have students compile the information into the chart on page 53.

Even though the handout uses the word *best*, discuss with your students that there really is not actually a best animal. They are looking for an animal that moves with ease through the school yard, finds food with no trouble, and faces few predators. That animal may be the one that has adapted over generations to survive in your local area.

Assessment/Next Steps: Evaluate students' charts for reasonable data collection. After compiling their individual chart information, students can work in groups with others who picked the same animal to create presentations for the class. Encourage the use of computer slide shows, posters, or even role-playing as each student attempts to persuade others in the class that his or her animal is best suited for life in your school yard.

Animals Living on the School Grounds

Name: _____ **Date:** _____

Directions

Find four animals to study. Use this chart to get started.

Animal Characteristics or Adaptations	Animal A _____	Animal B _____	Animal C _____	Animal D _____
Lives in trees				
Able to fly				
Exoskeleton				
Metamorphosis				
Lays eggs				
Cares for young				
Migrates				
Predatory				
Eats only plants				
Humans provide its food				
Swims				
Number of legs				

1. Refer to the chart to sum up what you have learned about animals that live in your area. What do they all have in common? Why? _____

   _____

   _____

2. Which animal do you think is best suited to life in your school yard? Why?

   _____

   _____

Put It All Together

Divide into groups. Students in each group decide which animal is best adapted for life in the school yard. Within your group, work on a presentation for the class. Convince other students that your animal is the best fit for life in your school yard. After the presentations have been given, conduct a survey to see if your classmates have changed their opinions.

School Yard Food Chain

Teaching Objective: to identify a food chain in the local environment

Why/How to Use This Lesson: This is a good culminating exercise for any unit on animals. If you have conducted several activities in this chapter, the food chain chart information can be taken from information that your students have already gathered.

Materials: handout, clipboard, pencil, binoculars (optional), hand lenses (optional)

Procedures and Tips: For outdoor observation prior to filling out the food chain chart, students may use binoculars to get a good look at animals in trees or use a hand lens to look at smaller creatures under rocks and in crevices.

If students cannot find animals, they should find evidence of their existence. Explain to the students that for them to use animals in the chart, they need to see at least one of the animals in the chain or at least see evidence of its existence. Some examples of a school yard food chain might be as follows:

1. parsley, caterpillar, spider, bird

2. wild onions, crickets, lizard, box turtle

Your students will surely discover many other examples. Check to make sure that student choices are conceivable for the environment on your school grounds.

Assessment/Next Steps: Evaluate the food chain for accuracy. In short, does it have a producer and at least three consumers? As an extension, have students use large sheets of drawing paper or bulletin board paper to complete food webs using their school yard food chains.

School Yard Food Chain

Name: _____ **Date:** _____

A *food chain* represents how energy is passed from plants to animals. In the chain, some animals (called *herbivores*) feed on plants. Other animals (called *carnivores*) feed on other animals. Some animals (called *omnivores*) feed on plants and other small animals. One animal eats another to survive. For example, a food chain in a pond might include a fish that eats algae, a turtle that eats the fish, and a raccoon that eats the turtle. The sun supplies energy to help new algae grow. Then the chain starts all over again.

Directions

Create a diagram in the space below to show a food chain you have seen in the school yard. Draw pictures and arrows to show the links in the food chain. Include one plant and three animals. Be sure you have seen at least one of the animals or at least be able to prove there is evidence of that animal living in the area. Label the food chain parts as either producers or consumers. Label the consumers as carnivores, herbivores, or omnivores.

Your Food Chain

3

Tag, You're It!

Teaching Objectives: to practice methods used by scientists to track migration of monarch butterflies; to participate in genuine scientific research as part of the international scientific community

Why/How to Use This Lesson: In the study of life science, this lesson fits into units on regulations and behavior or populations and ecosystems (NRC 1996). The activity allows teachers to illustrate for students how scientists gather information on the behavior of wild animals in context rather than through simple discussion or reading.

Materials: monarch butterflies (raised from caterpillars or captured wild), tags from Monarch Watch, butterfly nets, containers for butterflies, handout, clipboard, internet access (visit *www.monarchwatch.org*)

Procedures and Tips: Visit the Monarch Watch website well in advance of this lesson to order tags. If you do not have internet access, you can call Monarch Watch (1-888-TAGGING) for information on ordering tags. This is a fall activity, so plan to conduct this lesson between August and October. Once you receive the tags, you may wish to practice tagging a live butterfly on your own before you do so with your class. You may also consider doing this as a demonstration. Students should research the Monarch Watch website thoroughly before attempting this activity.

In addition to the instructions online, you will also receive tagging instructions with your tags. The main tip to emphasize for students is to be gentle with the butterflies. If you capture the butterflies wild, they tend to be far more active than if you raise them from caterpillars. Therefore, you may want to consider raising caterpillars if this is your first time working with monarchs. (Note to schools located west of the Rocky Mountains: Contact the Oregon Department of Agriculture, 635 Capitol St. NE, Salem, OR 97310, for information on tagging the western population of monarchs. Monarch Watch only tags monarchs located east of the Rockies.)

Assessment/Next Steps: Answers on the student handout allow teachers to assess understanding of the concepts in the lesson. After this lesson, students may engage in further studies of migrating animals through websites such as Monarch Watch or Journey North (*www.learner. org/jnorth*).

Tag, You're It!

Name: _____ **Date:** _____

You may have heard of scientists tagging birds or other animals. But have you ever heard of tagging an insect? The tagging of monarch butterflies takes place every year, and you can help. You can raise monarchs from caterpillars that your teacher provides, or you can locate milkweed plants and use a net to catch the monarch butterflies that come to feed there. For complete tagging instructions, look on the Monarch Watch website (*www.monarchwatch.org*) and study the pictures.

Directions
Think about the questions as you tag your butterfly. Record your answers below.

1. Why would scientists want to tag butterflies?
   _____
   _____

2. What problems might there be in designing a tag for such a small animal?
   _____
   _____

3. How is the information from a tagged butterfly used?
   _____
   _____

4. How can we help the monarchs along their journey?
   _____
   _____

5. How can the Canadian, U.S., and Mexican governments work together to help the migration of the monarch butterfly?
   _____
   _____

Helpful Hints
Tagging a monarch butterfly is tricky, to say the least! To tag it, you have to touch the butterfly's wings. This has to be done carefully so that the butterfly is not hurt. The trick is to hold the butterfly from the underside. This will keep the scales on the wings from rubbing off on your fingers. Be gentle. Make sure the tag is in the middle of the lower wing. The tag will not hurt the butterfly or keep it from flying.

Tag, You're It!

Name: _____ **Date:** _____

Show What You Know

Use this activity to spark your imagination! Write a newspaper article announcing the coming of monarchs to your community. Your teacher can even help you send the article to the editor of your local newspaper.

Read the information about butterflies. What butterflies might visit your school grounds? How can you find out what each type of butterfly likes to eat or drink? You can use exploration and experimentation to discover a butterfly's food preferences.

Think about what you would do if you wanted to find out (without asking them) if your friends like hamburgers or pizza best. How would you conduct the experiment? You might buy some hamburgers and a pizza. You could put everything on a table and see which food your friends choose to eat. In the same way, we can "set a table" for caterpillars and butterflies. Think about how you might do this. To gather the most accurate information possible, you would need to repeat the experiment several times. Repeating an experiment is good scientific practice and would take into account your friends who like both foods.

Directions

Design an experiment that answers the question "What do butterflies like to eat?" Use the scientific method: Select the type of butterflies that is native to your area. Form a hypothesis. Research and plan your experiment. Record your observations. Analyze your data and draw a conclusion on the basis of your results.

What Do Butterflies Eat?

Butterflies have one of nature's unique adaptations: metamorphosis. This means they completely change in form at different stages of their lives.

One of the most important results of metamorphosis relates to eating and food sources. Because of metamorphosis, adult butterflies do not compete for the same food sources as their young.

How Does This Work?

Young butterflies are larvae, but we usually call them caterpillars. Caterpillars eat leaves. Each species of butterfly has a special "host" plant. The female butterfly lays her eggs only on the leaves of this special plant, so when the caterpillars hatch and need to eat, their food source is right there for them.

Adult butterflies do not eat solids; they only drink their food. Sometimes the adult butterflies will puddle or gather on moist, sandy soil to drink water. More often, each butterfly drinks nectar through a proboscis, the slender "straw" that unfurls from its head.

It All Adds Up!

Math + Science + Outdoors = Fun

Creative mathematics teachers will tell you that math is everywhere and you cannot teach science without math. When you take students outside, you will find that math really is everywhere! Whether it is a geometry lesson using the angles of tree branches or a probability lesson about how many bird eggs will survive to adulthood, math plus science in the outdoors is a winning equation.

The national mathematics standards published by the National Council of Teachers of Mathematics (NCTM 2000, p. 76) call for students in the earliest grades to begin studying algebra in the form of patterns. What better place than nature to find patterns? Find patterns that occur in the outdoors, or create a space, such as the math patios described in Chapter 1, where students can make patterns and practice skills such as graphing, scale drawing, and keeping a calendar.

Many of the lessons in this chapter are for math patios, but there are also alternatives that might work more easily for you in your school yard setting. Some of these alternatives are:

- String and stakes. Use small wooden stakes and tie string around them to make a grid. If you have plastic tent stakes available, those would be a good alternative. This setup is one that could be used for a couple of weeks and then dismantled easily.

- Plywood and tape. If you cannot have a permanent patio outdoors, make a portable one. Measure equal squares on plywood and mark them off with masking tape or duct tape. Store the board when not in use. You can cut the plywood into smaller sections for easy transport to and from school grounds.

- Paint on a sidewalk or parking lot. If your school yard is limited in green spaces but heavy on the concrete, mark your grid with tape and then paint it onto the concrete with heavy-duty patio paint. Obviously, you will need permission for this before you start.

- Jump ropes laid in patterns. If you have no financial resources for paint, plywood, or string, you could lay jump ropes in a pattern for a single lesson. Any type of rope or yarn will work for a one-time use. (Safety note: Rope can be a tripping hazard. Make sure it is secured to the ground.)

- Indoor tile floor. If you cannot take the lesson outdoors, use the tiles on your classroom floor if you have them. Move the desks out of the way and mark the perimeter of your grid.

Each teacher's situation is unique. It takes cooperation among colleagues, administrators, students, and parents to build a permanent outdoor classroom. But one of these alternative ideas will still allow you to put the lessons in this chapter into action for your class and share the fun of learning on a math grid.

There is a time and place to break away from traditional presentations of mathematics and make math fun. Let that happen in your outdoor classroom learning lab or outside on school grounds. Integrate math into your outdoor science lessons. It will add up to success for both you and your students!

There are several ways you can use the lessons in this chapter. You may wish to use some lessons to meet a math objective, while others can be used as an extension of science or social studies lessons. Link science content to math concepts by explaining how various scientists use math skills in their work. On some lessons, the type of scientist who uses the skill from the activity is listed along with the objectives. Handouts for the activities follow this section.

Other Related NSTA Press Resources

Activities Linking Science With Math, K–4, by John Eichinger

Activities Linking Science With Math, 5–8, by John Eichinger

Stop Faking It! Math, by William C. Robertson

Don't Forget!

- The outdoor classroom is not an outdoor eating area. Eating in the outdoor classroom can attract unwanted wildlife.

- Wild animals should not be kept as pets in classrooms or homes. Also, be aware that animals such as butterflies ordered from a science supplier should not be released into the wild in areas where they are not native.

- Get advice specific to your location from a botanist, extension agent, or master gardener.

- Teachers should be aware of whether their state requires an Integrated Pest Management, or IPM, program so students will not be exposed to pesticides, herbicides, and other hazardous chemicals when working out-of-doors.

For a full list of safety tips, see pages 15 through 16.

Hunting for Numbers

Teaching Objectives: to identify a variety of numbers in nature; to practice basic mathematics operations

Why/How to Use This Lesson: Think about conducting this activity close to the beginning of the school year as an introduction to the outdoor classroom or school yard. This activity can also be a formative assessment activity for teachers who want to make informal observations of their students' basic skills in math.

Scientists Who Use These Skills: biologists

Materials: handout, clipboard, pencil, paper or plastic grocery bags

Procedures and Tips: This activity is an old-fashioned scavenger hunt in which students practice percentages, addition, and subtraction. Give each student a plastic grocery bag and the handout. Tell them they are going to hunt for the items on the handout, and ask them to pre-dict how many they will find. Ask them to put each item they find into their grocery bags.

If you are certain that several of the items listed on the handout cannot be found in your area, you may want to modify the chart. But leave those items if you want to let that fact work into the percentages the students will predict. This would also help them think about the environmental factors and seasonal changes that may influence which items are present and which are not.

Make sure that students return all of their natural items to nature when they finish the activity.

Assessment/Next Steps: The teacher will evaluate performance as each student or group brings items to be checked against the list before returning items to nature. You can incorporate additional math skills by including measurement. For instance, students collect two sticks that have a 10 cm difference in length. If a scale is available, students could collect rocks that weigh less than a certain amount.

Hunting for Numbers

Name: _____ **Date:** _____

Directions

You are on a hunt for the items you see below. Mark the square in the chart with a check mark when you have found the set of items. Put the items in your grocery bag. If you find everything, bring the bag to your teacher to be checked. When you are finished, answer the questions below the chart.

Before you start hunting, read through the whole chart and predict how many items you think you can find.

Prediction

I will find _____ of the 20 sets of items.

3 acorns	7 sticks	4 rocks	2 leaves from same tree	10 flower petals
1 ivy leaf	6 brown leaves	5 weeds	8 different leaves	2 seeds
4 pinecones	1 maple seed	2 oak leaves	6 dandelions	1 wildflower
9 pieces of bark	4 feathers	1 snail shell	12 blades of grass	11 pine needles

1. Add together the numbers in each vertical column. Add all of the numbers, including those items you did not find. Write your answers below the columns.
2. On the back of this paper, write five challenging word problems using these numbers and items for friends to solve.
3. What percentage of the total items (20) did you find? Count each set of items as one for this question.
4. Work with a partner. What is the difference in the percentage of items that each of you found? If you combined your charts, what would the new percentage of found items be?

Graphing Animal Behavior

Teaching Objectives: to gather data to include in a bar graph; to use science and math skills together; to observe animal behavior in nature

Why/How to Use This Lesson: Part of the aim of this lesson is to help students see the relationship between science and math. Consider supporting math instruction on graphing by showing how graphs provide evidence in scientific endeavors. Having students complete both written graphs and acting out human graphs will address different learning styles.

Scientists Who Use These Skills: biologists

Materials: concrete block patio, animals in natural setting, handout, pencil

Procedures and Tips: Students will produce both a written graph and a "human bar graph" using the data they have gathered from their scientific observations of school yard animals.

Take time to scout the school yard for animal activities. Birds will likely be the most easily observed animals in many places. To increase their activity, provide a number of special treats when you plan to conduct this activity. Temporary feeders can be created from clay flowerpot saucers filled with seeds or gallon milk jugs cut with a large hole and filled with seeds. Put one of these on a stump or near a tree. The feeders will probably also attract squirrels, which can then be observed as well, or you can observe insects. The number of times a bee or butterfly visits a flower will create good data for a graph.

Allow your students to gather data for about 30 minutes each day for a week. Then ask them to draw a graph in the box provided on the handout. Students should label their graphs with numbers and descriptors of the animal behavior measured. Ask to see each completed graph to make sure data are recorded correctly.

Assessment/Next Steps: After each student has a graph, have students take turns organizing their classmates to make the human graph. Help them come up with a method to make the graph workable. For example, each person in a line could represent five times that birds came to a certain bird feeder. Evaluate student performance based on the accurate representation of information in graphs. To take it further, additional graphs may be produced using computer programs or graphing calculators.

Graphing Animal Behavior

Name: _____ **Date:** _____

Directions

On your hundreds chart patio, or a grid of 10 in. × 10 in. squares, graph the data about an animal's behavior. You can use the data from other charts you have completed, or you can collect new data. The amount or type of food a butterfly or bird eats is easy to graph.

Getting Started

First, form a hypothesis—a prediction that can be tested with experiments or observations. For example, your hypothesis might be "Hummingbirds like nectar from flowers better than sugar water from a hummingbird feeder." Watch the animal over a period of time and record your data. Here is a sample of what your chart might look like:

Sample Data Gathered

Day	Number of times hummingbirds come to flowers	Number of times hummingbirds come to feeder
Monday	3	4
Tuesday	1	1
Wednesday	2	3
Thursday	0	6
Friday	1	5
Total	**7**	**19**

Graph It!

Use your data and turn your patio into a "human bar graph." If you were using the sample above, make two columns for Monday. Ask three classmates to stand in Column 1, and four classmates to stand in Column 2. By the time you place students in a column for every day and the preferred food source, the whole class will be involved!

More Graphing!

Gather paper, a pencil, and a ruler. Using the data you collected during this experiment, make a bar graph on paper to represent your results. Label the graph clearly and give it an appropriate title.

What's Your Net Worth?
Measuring and Using an Insect Net

Teaching Objectives: to practice measurement skills; to design an experiment to test the effectiveness of an insect net

Why/How to Use This Lesson: In an integrated unit about butterflies, this activity supports math instruction on measurement and encourages students to try experimental design.

Topic: Systems of Measurement

Go to: www.scilinks.org

Code: OS010

Materials: wooden dowel, duct tape, wire clothes hanger, wire snips, roll of fabric mesh, fishing line, tape measure, large dull needle, heavy thread, handout, clipboard, pencil

Procedures and Tips: Your students can build a butterfly net with simple materials from home or a hardware store. Then they can measure their own net and compare it to other students' nets. Make sure students make their nets with dowels of different lengths and nets of different sizes to help with comparisons.

Follow this procedure for making an insect net:

1. Cut the hook off the clothes hanger and untwist the top.

2. Shape the wire into a circle, leaving enough straight wire to go alongside the dowel about 25 cm.

3. Wrap duct tape around the wire and dowel to secure the wire circle.

4. Measure how much fabric mesh you will need to go around the wire circle.

5. Secure the fabric to the wire circle by tying it with fishing line. It can be tied in small knots or sewn all the way around and down the open edge with the needle and thread.

Assessment/Next Steps: After the nets are complete, your class should take the measurements shown on the student handout. As an additional math activity, students can calculate the cost of their nets if you provide them with a list of prices based on receipts for the supplies. Allow students to use their nets to catch butterflies and other flying insects, then gather information on the size of each net built in the class and the number of insects caught by that net. Find out if there are certain patterns of success that relate to the sizes of the nets. Finally, have students graph the results. Assess student performance based on accurate representation of data in graphs.

What's Your Net Worth?
Measuring and Using an Insect Net

4

Name: _____ Date: _____

Directions
Answer the questions about your insect net. You will need a meterstick.

Measurements
1. The length of my net's handle is _____ cm.

2. The circumference (distance around the opening) of my net's hoop is _____ cm.

3. The diameter (distance across the opening) of my net's hoop is _____ cm.

4. The depth of the actual net from the hoop to the bottom of the inside is _____ cm.

Success Rate
5. While using the net outdoors, I attempted to catch _____ insects; I actually caught _____ of them.

6. The percentage of those I caught was _____.

Compare and Contrast
After collecting insects, work with a partner to answer the questions.

7. Whose net caught more butterflies or other insects?
   _____

8. Why do you think this net caught more?

9. On another sheet of paper, make a bar graph to compare the size and results of the two nets. Graph at least four sets of numbers. Then make a conclusion about what makes a good insect net on the basis of your results.

67

4

What Can You Learn From a Seed?

Teaching Objectives: to practice measurement and percentages; to germinate a seed and make quantitative observations

Why/How to Use This Lesson: Students in elementary and middle school benefit from lessons involving seeds as they learn about the characteristics of organisms, life cycles of organisms, and regulation and behavior (NRC 1996). As students move through upper elementary grades into middle school, they can begin to understand the different parts of seeds and distinguish between the needs of seeds versus the needs of mature plants. This lesson would easily integrate mathematics near the beginning of a unit on plants or life cycles in general.

Materials: seeds or dried beans, paper towels, resealable plastic bags, water, metric ruler, digital scale or balance scale, handout, container and potting soil or garden space (optional), "Needs of Seeds" assessment probe (Keeley, Eberle, and Tugel 2007))

Procedures and Tips: Growing seeds is a common science activity in many classrooms, but it is also an effective mathematics lesson. Consider starting with the formative assessment probe listed above.

Dried beans from the grocery store are easy to use for this activity because they are big and easy for children to handle, weigh, and measure.

If you prefer to grow flowers, buy sunflower seeds. For a larger vegetable seed, think about planting pumpkins.

Make sure you have metric rulers marked in millimeters and centimeters. Weighing the seeds may be a little trickier than measuring length, depending on your scale. A high-quality digital scale works best, but nearly any scale from a science kit will do. If one seed does not weigh enough to make the scale register a number, ask the students what they should do. With encouragement, they will come up with the idea of weighing the entire package of seeds and dividing the weight by the number in the bag.

Students will germinate the seeds in plastic bags. Ask them to moisten their paper towels lightly after folding the seed gently inside and place the towel in the resealable bag. Keep the bags near a window, but not in direct sunlight. As the days pass, watch the plants sprout, and follow the questions on the handout with your students.

Assessment/Next Steps: Evaluate students on the accuracy of their measurements and reasoning for deciding how to determine the mass of the seeds. When the seedlings are getting close to outgrowing the bag, you may wish to provide a planting space outdoors or in a container. Otherwise, send the seedlings home and ask students to plant them in their own yards or gardens. This is also a good way to get parents involved in math and science instruction!

What Can You Learn From a Seed?

Name: _____   **Date:** _____

Have you ever planted a seed? If so, what kind of seed? What did you learn from it? Get a seed from your teacher. Examine it carefully. Predict how the seed will change. You will be amazed at all that you can learn from one tiny seed!

Day 1

1. The weight of my seed is _____ g.

   If my seed were too light to weigh, I would calculate the weight by

   _____.

2. The length of my seed is _____ mm.

3. I predict my seed will become a _____.
   (type of plant)

To be able to measure your seed again, you will start to grow it on a damp paper towel inside a plastic sandwich bag. Your teacher will show you how to do this. You will measure it again in three days. Predict how much you think it will grow by then.

4. I think my seed will weigh _____ g and be _____ cm long in 3 days.

Day 4

1. My seed (and the plant, if it is started) weighs _____ g. The difference between this weight and my prediction is _____ g.

2. My seed (and plant) is _____ cm long. The difference between this length and my prediction is _____ cm.

3. Find the percentage of growth that took place. For example, if you predicted your seed would weigh 10 g and it weighs 9 g, then 90% of the growth in weight that you predicted took place. Figure an answer for both weight and length. Average the two percentages.

4. Work in a group of three students. How do the actual results compare with each person's predictions? Were they close? Make a bar graph to compare the predictions and actual measurements. Choose a speaker from your group to share your results with the class.

Follow-Up

Plant your seeds in a pot or outdoors in a garden. You can make more predictions and see what happens to your plant. After your seedling is planted in dirt, measure the height from the ground to the top of the plant. Do this every day for 14 days. Make a line graph to represent the data you collected.

Germination Determination!
Old Seeds, New Seeds

Teaching Objectives: to compare and contrast the germination rates of seeds produced for different years; to create a line graph based on data gathered in an experiment

Why/How to Use This Lesson: Use this activity to allow students to study seeds over time, perhaps with the previous activity as part of unit on plants or life cycles. The needs of mature plants will be distinguished from the needs of seeds, so this is definitely a lesson for upper elementary or middle school. If adapting for lower grades, consider modifications according to your knowledge of your students' needs.

SCi LINKS.
THE WORLD'S A CLICK AWAY

Topic: Seed
Germination

Go to: *www.scilinks.org*

Code: OS011

Materials: two or more packets of seeds for the same plant, each produced during a different year; outdoor garden spot or pots and potting soil for indoors; water source for plants; pencil; handout; clipboard; craft sticks for plant markers

Procedures and Tips: Many stores will give teachers seeds at the end of the summer-growing season. If you can collect these packets about the time the school year starts, you can buy new seeds in the spring and have two years' worth of seeds for this experiment.

Follow this procedure for the activity:

1. Plant the seeds from two different years under identical circumstances. For ease in mathematics calculations, choose 10 seeds from each year's packet to plant.

2. If planting in pots, use identical pots and measure the amount of potting soil so it is equal.

3. Plant the seeds according to package instructions, usually just below the surface. This can be done as a whole-class activity with only two flowerpots of seeds, one from each year's seed packet.

4. Label the pots with plant markers to show the type of plant and year of seeds.

5. Students make predictions and gather data using the handouts.

If you need a good source of seeds, apply for a grant at *www.kidsgardening.com* and you will receive a variety of seeds in the mail.

Assessment/Next Steps: Evaluate the students' accuracy as they fill in the chart as the seeds grow. As a culminating activity, on another sheet of paper students may create a line graph comparing the seeds from two different years.

Germination Determination!
Old Seeds, New Seeds

Name: _____ **Date:** _____

Have you ever read a seed package carefully? Have you noticed that seed packages have a date stamped on them? The date is the year during which the seeds should be planted. Some people will plant seeds that are more than a year old, and other people discard outdated seeds.

Directions

Germination—the growth of a seedling from a seed—can be tested in this simple experiment. Work in pairs. One person plants the older seeds, and the other person plants the newer seeds. As your plants grow, fill in the chart below or your own chart on another sheet of paper. Make a conclusion about your results and answer the questions.

Date	Quantitative Data	Qualitative Data (Comments/Observations)
Example Planted 09-01-12	2010 Sunflowers/2011 Sunflowers	Seeds look the same regardless of year produced
Example 09-08-12	2010 Plants 6 cm/2011 plants 7 cm	No difference except in height—appear the same

Quantitative data are noted in numbers such as weight, height, and length.

1. What happened to your seeds during the first week? How did they change?

    _____

    _____

2. Explain the difference in the growth of plants from new seeds and older ones.

    _____

    _____

3. Think about the result. Would you plant old seeds during the next year? Why or why not?

    _____

    _____

Too Many Seedlings
Population Density

Teaching Objectives: to predict the outcome of too many seeds sprouting in a small area; to understand the effects of overpopulation on other species

Why/How to Use This Lesson: Use this activity in a study of regulation and behavior (NRC 1996) or an integrated unit on plants or the needs of organisms (including space). This may also be used in guiding students through the process of spacing out plants when creating an outdoor classroom.

SCiLINKS.
THE WORLD'S A CLICK AWAY

Topic: Plant Growth
Go to: *www.scilinks.org*
Code: OS012

Materials: seeds, soil, planting container, water (or access to dense seedlings outdoors), handout

Procedures and Tips: If you do not have any seeds, try finding some outdoors. A pinecone or dandelion bloom may provide enough seeds to conduct this activity.

Follow this procedure for the activity: If possible, find a naturally occurring set of seedlings that have sprouted very close together. Look for seed-bearing items such pinecones, seedpods, or dead flowers that have fallen off the plant. These will sometimes dump all of their seeds in one spot. If you can find this on your school grounds, take your class to make observations. Use the student handout questions to get students thinking. Have some seeds, soil, and a pot available. Guide students to a discussion about experimenting with planting seeds close together to see what happens. As a comparison, use another flowerpot to plant the seeds in the recommended space. Students should make observations over a period of days and weeks to see what happens to the seeds and subsequent seedlings.

Assessment/Next Steps: Evaluate students' thought processes in their responses to prompts on the handout. Refer back to this activity throughout the school year when planting either in pots or outdoors as a reminder to students regarding each plant's need for space to grow.

Too Many Seedlings
Population Density

Name: _____ **Date:** _____

Have you ever been on a crowded elevator? How long is it comfortable to be so close together? Have you ever thought about why animals need their own "territory" or space? What about plants? Have you ever seen dozens of seedlings growing in a small area? How can you experiment to determine what happens to plants that are overpopulated in a small area. What if all the seeds in a pinecone fell out of the cone and straight to the ground—all sprouting within a few centimeters? What about an entire seedpod?

If possible, look outdoors for a dense area of seedlings. Try to determine the species of plant and document with measurements, sketches, and digital photographs. Whether such an area of plants can be found or not, write a few sentences about what would happen if a plant along the edge of the cluster were removed and planted alone with more space around it. How would it compare to those left together in a week or month?

How can you replicate this scenario in a flowerpot or similar container? Design an experiment to compare a dense population of seeds to a sparse population of seeds. Write your steps below. Work with your teacher to determine what materials are available.

Take It Further
Interview a gardener, farmer, or botanist. Find out what steps are taken when planting to space plants properly.

Weather or Not

Teaching Objectives: to average a group of numbers; to make a scientific prediction

Why/How to Use This Lesson: In elementary school, students should study changes in the Earth and sky (NRC 1996). In middle school, they should study Earth systems, including how global atmospheric patterns influence local weather (NRC 1996). In this lesson, students will gather data about local weather and, depending on their grade level, make appropriate inferences regarding weather patterns.

Scientists Who Use These Skills: meteorologists

Materials: nonmercury outdoor thermometer, rain gauge (available from most garden centers), handout

Procedures and Tips: This lesson will help students recognize that meteorologists are scientists who use math skills. You may wish to invite a weather forecaster from a local television station to speak to your students. Opening this lesson with a real meteorologist, whether recorded or live, can help your students think the way a scientist thinks. If that is not possible, look for a video on weather, or tape a short segment from a weather forecast to share with your students as an introduction.

Students will watch weather forecasts to gather data on weather predictions for the week, but they will record the air temperature and precipitation measurements at school. The best place to put a thermometer and rain gauge is in a grassy area at least 2 m from the building or concrete surfaces. This will help students avoid having the air temperature affected by the transfer of heat energy from these surfaces.

Assessment/Next Steps: Evaluate how students complete the chart and graph the information according to instructions on the handout. As a culminating activity, you may want to use a video camera and have students record their own weather show complete with data from their experiments and predictions for the coming days. Your school may allow you to show this on a closed-circuit television system if it is available.

SCI LINKS.
THE WORLD'S A CLICK AWAY

Topic: How Can
Weather Be Predicted?

Go to: *www.scilinks.org*

Code: OS013

Weather or Not

Name: _____ Date: _____

How do we talk about weather? Do we use only words to describe weather, or does it take more? Of course it does! Numbers help us measure the weather and the way it affects us. In this activity, you will take measurements and make observations just like a meteorologist (a scientist who studies atmospheric conditions and makes weather predictions).

Materials

nonmercury thermometer, rain gauge, weather reports (television, newspaper, or internet)

Directions

Watch a television weather forecast or read newspaper or internet weather reports. Look at what high air temperatures and rainfall are predicted for the week. Using everything you have learned, make your own prediction about the high temperature and the amount of rainfall for each day. Record your predictions in the chart below. To make measurements for the second week of this activity, place a thermometer and rain gauge on your school grounds. If this is not possible, read or listen to weather reports to get your data.

	Monday	Tuesday	Wednesday	Thursday	Friday
Forecast for Week 1 **temp/rain**	(Ex. 60/1 inch)				
Your prediction for Week 2					
Actual numbers for Week 2					

Analyze the Data

1. What was the average difference between your prediction and the actual air temperature?

2. What is the relationship between the amount of rain and the temperature? Use math concepts to explain your theory.

3. What was the total rainfall amount for Week 2? What total did you predict?

4. On the other side of this sheet, write a word problem that requires a prediction for another week, based on your data from this week. Copy your word problem onto another piece of paper and trade with a classmate. Check each other's work and help make any corrections needed.

Math in a Tree

Teaching Objective: to practice measuring angles, circumference, and diameter

Why/How to Use This Lesson: This activity could be used as an introduction to measurement, a midpoint practice, or a culminating activity after a unit on measurement is under way.

Scientists Who Use These Skills: arborists, foresters

Materials: tape measure, meterstick or ruler, clear protractor, handout, clipboard, pencil

Procedures and Tips: You may want to scout the school yard or a nearby park, if necessary, to find a tree that will work for this activity. Students will need to reach leaves on low-hanging limbs or, if the season is right, be able to pick up leaves that have fallen on the ground.

The questions on the handout need to have clear answers, so the tree you choose will need to have a circumference that can be measured using tape measures. The tree needs to have a number of branches that allow students to identify angles visually. Before teaching this lesson, students will need to know these terms: circumference, diameter, triangle, angle, and protractor.

Assessment/Next Steps: Check student work on the handout for reasonable answers and correct computations. Challenge students to write a math word problem using some information they have gathered about trees in your school yard. The GLOBE program (*www. globe.gov*) uses information students gather about trees as part of its international study of the natural world. For enrichment, invite a professional forester to speak to your class, highlighting how he or she uses mathematics and science in his or her career.

Math in a Tree

Name: _____ **Date:** _____

Directions
Math is everywhere in the great outdoors. You can add it all up by finding the sum of math in a tree. First, choose a tree on your school grounds or in a nearby park. Then answer the following questions about that tree.

Materials
flexible tape measure, meterstick or ruler, clear protractor

1. What is the circumference of the trunk 20 cm up from the ground? 50 cm up? Next, lay your tape measure on the ground in a circle to show each measurement. What is the diameter of each measurement?

   20 cm up   Circumference = _____ cm   Diameter = _____ cm

   50 cm up   Circumference = _____ cm   Diameter = _____ cm

2. Find a branch that you can either reach or see clearly. Look through your protractor at the branch. Measure the angle between the branch and the trunk of the tree. Repeat this step with another branch. If these two angles were part of a triangle, what would the measurement of the third angle in the triangle be?

   Tree angle #1 = _____ Tree angle #2 = _____

   Remaining angle needed to complete a triangle: _____

3. Find four different-size leaves from your tree. Find the average lengths and widths. What is the average size of a leaf from your tree?

   Average Length = _____ cm   Average Width = _____ cm

4. Look carefully at your tree. Back up a little from it. Estimate the height of your tree.

   I estimate the tree is _____ m tall.

Building and Using a Compost Bin

Teaching Objectives: to practice using measurement skills; to learn to read a thermometer accurately; to study decomposition of organic materials

Why/How to Use This Lesson: As students move through grades 5 through 8, they need to use quantitative data to support qualitative data (NRC 1996), and a compost bin will give them the opportunity to gather this quantitative data. Using a compost bin may also give students the opportunity to explore the process of decomposition.

Topic: Composting
Go to: *www.scilinks.org*
Code: OS014

Scientists, Who Use These Skills:

botanists, food scientists

Materials for Building: wire fencing (hog wire or chicken wire), six pieces of treated lumber (2 in. × 4 in. × 8 ft.), galvanized nails, hammer, hand saw, heavy-duty carpenter's stapler, safety goggles, gloves, piece of plywood (optional)

Materials for Activities: nonmercury thermometers, food scraps, yard scraps, handout, clipboard, pencil

Procedures and Tips: Build a compost bin with your students, and use it to recycle fruit and vegetable scraps and yard waste into valuable compost for your school garden.

To make the bins: Cut each board into two 4 ft. lengths. Use nails to fasten the 12 boards into a 4 ft. × 4 ft. × 4 ft. frame. Use a carpenter's stapler to affix the wire fencing to the four vertical sides of the cube. You may leave the top open or use a piece of plywood as a cover. (If the compost bin is uncovered, it will work, but be sure to check your local ordinances to see if a cover is required by law.)

Place the bin about 30 ft. (10 m) away from the school building if possible. The bin may attract insects. Some compost bins emit an odor, but you can avoid that by limiting the waste you put in it and stirring often. Use apple cores, banana peels, or scraps from a school salad bar. Never include animal by-products in your compost. Gardening scraps can be included (e.g., clippings from flowers and shrubs). Layers should alternate, as the food scraps will hasten the decomposition of the yard waste. Also, keep the materials in the compost bin damp by adding water whenever necessary.

One way to test the compost bin for effectiveness is to determine the temperature inside the heap. Some science classrooms are equipped with thermometers that are about 30 cm long, which your students can stick through the wire and into the heap as far as possible. The compost heap should show an internal temperature several degrees higher than the air temperature. Try taking the bin's temperature at different times during the day or every day for a week at the same time. These numbers can be represented on a graph drawn on the math patio blocks of your outdoor classroom.

Assessment/Next Steps: Evaluate students' ability to gather reliable data from the compost bin. If graphing calculators and probes are available, using a temperature probe to gather data would be an appropriate extension of this lesson.

Building and Using a Compost Bin

Name: _____ **Date:** _____

Directions

Is there math in a compost bin? You bet there is! Find out what's "cooking" inside. Use a nonmercury thermometer to find the difference between the temperature inside the compost bin and the air outside the bin. Complete the steps below and record your data.

1. Finish the prediction: I think the temperature in the compost bin is (lower than / higher than / the same) as the temperature outside.

2. Measure the temperature inside and outside the bin. Complete the chart below. Record the temperature every hour or at the same time every day. *Note:* When you measure the heat of the compost, push the thermometer into the compost heap as far as it will go.

Temperature Chart

Record the Temperature	Day/Hour 1	Day/Hour 2	Day/Hour 3	Day/Hour 4	Day/Hour 5
Inside the Compost Bin	(Example: Oct. 1 at 10:30 a.m.)				
Outside the Compost Bin	(Example: 29 degrees C)				

On another sheet of paper, answer the following questions:

3. Is there a pattern in the temperature readings? Make a graph to represent the data.

4. What were the results of your temperature experiment? Is the temperature inside the compost bin generally cooler or warmer than the outside air? Why?

5. Write a paragraph describing your findings and conclusion. Describe your understanding of the role that heat energy plays in decomposition.

6. What would you do differently if you did the experiment again?

Measuring and Analyzing Erosion

Teaching Objectives: to apply use of measurement skills; to recognize the effects of water erosion on land

Why/How to Use This Lesson: Consider integrating the study of Earth science into a study of local or school yard erosion. Search your social studies standards to learn how farmers tackle erosion problems, and use math standards on measurement. Integrating standards from various disciplines presents a powerful opportunity for service learning that can lead to reducing erosion in the school yard or another local, public space.

Topic: Water Erosion
Go to: *www.scilinks.org*
Code: OS015

Scientists Who Use These Skills: landscape architects

Materials: meterstick or measuring tapes, flag markers, clipboard, handouts, pencil, calculators

Procedures and Tips: This activity can be done indoors or outdoors, but if weather produces erosion in your school yard, that is the ideal location. Scout the yard for even the slightest slope, and look for evidence of erosion and deposition. Make sure your administrator knows what you are doing and approves of your students' observing the erosion in a particular place. If the administrator resists, offer to have students propose solutions to erosion problems on school grounds. You may wish to have different classes work in different areas of the school yard for comparison.

Once you obtain permission, students can gather data from the area of erosion. Follow this procedure:

1. Find an area where erosion has begun. Mark both sides of the gully with a flag. (You can purchase builders' mark flags at a hardware store.) Make sure the flags are firmly in the ground.

2. Measure the width from flag to flag by laying a meterstick straight across the gully.

3. Measure the depth by leaving the meterstick and measuring from the meterstick to the deepest spot in the gully. To measure deposition, place a ruler in the ground in the location where the soil is gathering.

4. Record your observations.

5. Repeat at regular intervals following rain showers.

6. Analyze your results.

Variation: If you need to do the activity indoors, use an under-the-bed plastic container filled with soil or sand. Put it on the floor with one end propped up on blocks or a brick. Pour a controlled amount of water on it each day for a week. Gather your results each day, and have students use the handout as with the outdoor version of the activity.

Assessment/Next Steps: Make sure that students complete the handouts with appropriate numbers. If students can make improvements in the school yard, consider having them share before and after photographs at a PTA meeting or through the local news media.

Measuring Erosion

Name: _____ **Date:** _____

Erosion—the movement of soil caused by water or wind—can be observed easily in the school yard. You can also measure deposition, which is the deposit of soil after it has eroded.

Directions
Follow the local weather and choose some days that rain is predicted. Be patient, and you will be able to observe the mighty force of erosion. Be sure to make accurate marks when measuring the ground with a meterstick or measuring tape and keep exact records. Fill in the blanks below.

Date of First Observation: _____

1. The original width of the eroded gully is _____ cm.

2. The original depth of the eroded gully is _____ cm.

3. An area of deposition is identified: Yes _____ No _____

Date of Second Observation: _____

1. Width _____ cm   2. Depth _____ cm   3. Deposition depth _____ cm

Date of Third Observation: _____

1. Width _____ cm   2. Depth _____ cm   3. Deposition depth _____ cm

Date of Fourth Observation: _____

1. Width _____ cm   2. Depth _____ cm   3. Deposition depth _____ cm

4. Draw a picture of the gully on the back of this page.

Analyzing Erosion Measurements

Name: _____ **Date:** _____

Directions

Now that you have gathered data on erosion, what does it mean? Can you draw conclusions about the erosion that you observed? You need to look at all of the information you have gathered. Put it into the graph below. You may use the grid below for a line graph. Be sure to label the graph properly. (If you have a math patio outdoors, you can draw your graph on the patio with sidewalk chalk.)

1. Look at your graph. What trend does it show in the erosion? Does the deposition change in the same way? Is it what you thought it would be? How did the amount of rain affect the erosion? Write a paragraph about your observations. Make sure your paragraph clearly explains the information on your graph.

   _____
   _____
   _____
   _____
   _____
   _____

2. Use your drawing of the gully from the activity Measuring Erosion. Add arrows and words to show how the area changed from the first time you observed it.

Don't Forget!

- Teachers should make sure that the outdoor classroom is not located near a playscape and also that the selected area was not previously the site of a wooden playscape. On a related note, when working with "treated timber," teachers should make sure it was fixed with non-arsenic preservatives.

- Make sure appropriate safety training and precautions are taken for all activities. When working outside, students should use appropriate personal protective equipment (PPE), including safety glasses or goggles, gloves, close-toed shoes, hat, long-sleeve shirt, pants, sunglasses, and sun screen.

- Students must use caution when working with sharp objects when working with timbers.

- Students should wash hands with soap and water after activities.

For a full list of safety tips, see pages 15 through 16.

Simple Machines Are for the Birds

Teaching Objectives: to design an experiment using simple machines; to practice using percentages

Why/How to Use This Lesson:

Students need to do more than memorize names of the different simple machines. Using this lesson will give students practice using simple machines to complete a task. By the time they are in middle grades, students need to understand how a force can act on an object (NRC 1996), and the use of simple machines will teach this lesson.

Scientists Who Use These Skills:

ornithologists, engineers

Materials: bird feeder, small buckets, seed, funnel, pulley, rope, boards, screws and nails, hammer, screwdriver, handout, pencil

Procedures and Tips: Students will use measurement skills to design an invention that fills a bird feeder with seeds. Start by distributing the handout and discuss with your students the different uses of simple machines in everyday life. A hammer, for instance, is both a lever and a wedge. A butter knife is a wedge. Ask them to think of other examples. Discuss which simple machines might work best for filling a bird feeder.

You can do this activity as a demonstration if you only have one bird feeder. That would

be the most cost-effective way to attempt this lesson. Position the funnel in the top of the bird feeder so that the seeds will have a fair chance of dropping into the feeder.

Give your students the freedom to try some things that do not work. That is what will yield the numbers your students will manipulate on the second part of the handout.

Whatever your students design, they should be able to handle the simple machine and let it pour seeds into the funnel and bird feeder. Discuss how one practical use of a successful machine would be a way to fill feeders that are out of reach. This would allow feeders to be placed higher up in trees, away from animals that prey on birds.

One simple machine that filled bird feeders successfully in the past was an inclined plane that took the seeds up above the feeder with a conveyor (the conveyor took additional materials that included lumber, a plastic handle, and rubber matting). Another successful machine (made with materials listed for this activity) included a pulley and rope lifting a bucket above the funnel and a second rope that pulled on the bucket to tip it.

Assessment/Next Steps: Evaluate students on their ability to discuss the merits of their designs and the use of quantitative data to support the designs. You may wish to have your students gather data for several designs so they can compare and contrast.

Simple Machines Are for the Birds

Name: _____ **Date:** _____

Use your measurement skills to design an invention that fills a bird feeder with seeds. This is a discovery activity. You will need to experiment with the materials to find a design that works. There is no right or wrong way to design your invention. Be creative!

Hypothesis
If I use one or more simple machines to invent a bird feeder filler, I can put 30 g of birdseed in a feeder using only my invention.

Materials
bird feeder, birdseed, funnel, pulley, rope, boards, wood screws or nail, hammer, screwdriver, small bucket, metric ruler, scale

Background
There are six types of simple machines: screw, pulley, inclined plane, wheel and axle, wedge, and lever. These machines are designed to make work easier for us. When two or more of these are combined, they become a complex machine. Using the materials you have, design an invention that makes it easier to fill a bird feeder.

Simple Machines
Look at the pictures of simple machines below. Think of two other examples for each category and write those names in the blanks. On the back of this paper, draw two examples for the wheel and axle category.

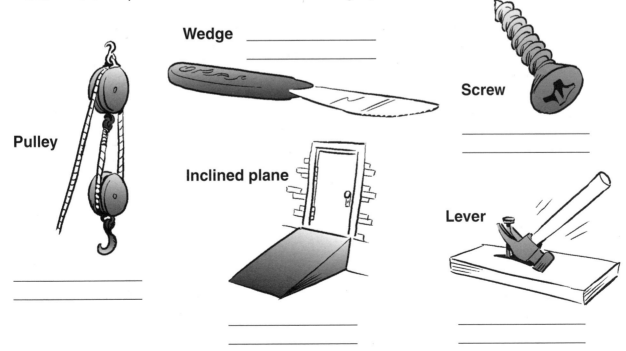

Wedge _____
_____

Screw
_____
_____

Pulley
_____
_____

Inclined plane
_____
_____

Lever
_____
_____

4

Simple Machines Are for the Birds

Name: _____ Date: _____

Directions

1. Make a sketch of how you plan to put your materials together to fill the feeder. Label each side of your machine with its measurement in centimeters. Use arrows and captions to show your invention in action.

2. Now you will test your invention. Answer the questions below after you have tested your invention three times and recorded your data.

Questions

1. What items did you need to measure (wood, rope, etc.) as you put your invention together?

_____

_____

2. How much birdseed (in grams) actually dropped into the bird feeder on each of your three tries?

1st_____ 2nd_____ 3rd_____

3. What percentage of the total amount of seeds made it into the bird feeder each time?

1st_____ 2nd_____ 3rd_____

4. Was the hypothesis correct? Were you able to fill the feeder by using simple machines?

_____

_____

5. What changes would you like to make to your bird feeder filler to improve its accuracy?

_____

_____

NATIONAL SCIENCE TEACHERS ASSOCIATION

Reading and Writing About Nature

The integration of science and language arts is an effective way for teachers to enhance students' comprehension and critical thinking and the requirements of their curriculum. Both writing and reading can be readily emphasized during science lessons. When students write about what they learn in science, the science concepts are reinforced, and students can practice writing skills that they will use across the curriculum. Many students love to learn about wildlife, so reading about wild animals and nature is an easy way to engage students in using reading strategies during science class periods.

When selecting writing assignments related to the outdoors, think about outdoor lessons that have sparked the most interest from your students. If possible, find a nature journal written by a scientist and read an example to your students. Some journals show how scientists and naturalists use text, sketches, and measurements to record sightings of unfamiliar plants or animals, which students often find fascinating.

Reading materials related to the outdoors are plentiful. A great book about an outdoor topic can be powerful to read aloud in your outdoor classroom. Sometimes you can follow this indoors with a clip from a movie. One of the best sources for teachers is the National Science Teachers Association and Children's Book

Council's annual list of books, Outstanding Science Trade Books for Children. These books have been reviewed by teachers and carefully screened for quality. You can find several years' worth of these lists on the NSTA website (*www.nsta.org*). Media specialists often order every book on this list, so check to see if your school has them.

There are many alternatives for engaging your students' imaginations and weaving together science and language arts activities. Here are just a few alternatives:

- Read several books about wetland environments, then set up an aquarium in the classroom to contain a wetland ecosystem. For example, you could have water plants, minnows, snails, and a small water turtle in your aquarium. Students can keep journals about their observations.

- Read an ecological mystery, such as *Who Really Killed Cock Robin?* Then set up a mystery in your outdoor classroom for your students to solve. "Plant" evidence, such as a bird nest on the ground, some feathers, or an empty turtle shell. Encourage students to write their own mystery story that connects and explains the "clues."

- After reading a migration story, such as *Fly Away Home*, by Patricia Hermes, locate the migration routes of Canada geese. Find a town with a school near one of the main routes. Use an internet directory to locate the school's address. Have your class write a letter to the students there, asking them to write to you when the Canada geese migrate to

their home. (The book *Fly Away Home* is a novelization of the movie by the same name, not to be confused with Eve Bunting's story.)

- Read an essay by John Muir or another naturalist, and then turn the essay into a play that your class can perform on Earth Day.

Enjoy reading and writing outdoors with your kids. Their enthusiasm might even inspire you to write your own creative science-related work!

Other Related NSTA Press Titles

Using Science Notebooks in Elementary Classrooms, by Michael P. Klentschy
How to … Write to Learn Science, 2nd edition, by Bob Tierney and John Dorroh
Picture-Perfect Science Lessons, by Karen Rohrich Ansberry and Emily Morgan
More Picture-Perfect Science Lessons, by Karen Rohrich Ansberry and Emily Morgan

A really good book makes the science-reading connection even stronger. Here are some suggestions for books in the classroom that your students will enjoy for both their wonderful writing and science content.

Reading Suggestions

This chapter is organized into three distinct sets of activities: Writing Basics—activities designed to help you extend your teaching of writing; Poetry in Nature—activities in which students are asked to think creatively about what they see outdoors; and Reading About Nature—activities that have students respond to literature selections. Handouts follow each activity.

Reading and Writing About Nature

Title	Author	Publisher	ISBN
Mathematics			
Ten Seeds	Ruth Brown	Knopf Random House	978-0-375-80697-1
Butterfly Count	Sneed B. Collard III	Holiday House	978-0-8234-1607-3
Paleontology/Archaeology			
Sabertooth	Patrick O'Brien	Henry Holt and Company	978-0-8050-7105-9
The Magic School Bus in the Time of the Dinosaurs	Joanna Cole	Scholastic	978-0-590-44689-1
Science Grades K–5			
Into the Woods: John James Audubon Lives His Dream	Robert Burleigh	Atheneum	978-0-689-83040-2
It's a Butterfly's Life	Irene Kelly	Holiday House	978-0-8234-1860-2
Here Is the Wetland	Madeleine Dunphy	Web of Life Children's Books	978-0-9773795-9-0
Predators	John Seidensticker and Susan Lumpkin	Simon & Schuster Books for Young Readers	978-1-4169-3863-7
Butternut Hollow Pond	Brian Heinz	Millbrook Press	978-0-8225-5993-1
Science Grades 4–8			
Animal Tracks and Signs: Track Over 400 Animals From Big Cats to Backyard Birds	Jinny Johnson	National Geographic Society	978-1-4263-0253-4
The Talking Earth	Jean Craighead George	HarperCollins	978-0-06-440212-5
I Want to Be an Environmentalist	Stephanie Maze	Harcourt	978-0-15-201939-6
Field Trips			
Field Trips: Bug Hunting, Animal Tracking, Bird -watching, Shore Walking	Jim Arnosky	HarperCollins	978-0-688-15172-0
National Audubon Society First Field Guide: Trees	Brian Cassie	Scholastic	978-0-590-05490-4
Girls Who Looked Under Rocks: The Lives of Six Pioneering Naturalists	Jeannine Atkins	Dawn Publications	978-1-58469-011-5

My Friend, My Bud

Teaching Objectives: to review the parts of speech (nouns, adjectives, action verbs, prepositions); to use figurative language (simile or metaphor); to examine the characteristics of plants

Why/How to Use This Lesson: Elementary students are ready to learn about the characteristics of living things (NRC 1996). Studying a bud is also useful in a unit on life cycles or seasons. Integration of science into language arts lessons makes it possible to give more time to science, which is often a challenge in the elementary classroom.

Topic: Plant Characteristics

Go to: *www.scilinks.org*

Code: OS019

Materials: handout, buds on cut stems or plants outdoors, pencil, journal or notebook

Procedures and Tips: Make a survey of the school grounds for buds on trees or shrubs. If you plan to do the lesson indoors, you can prune stems with buds from your own backyard. If you will use stems that you cut the evening before the lesson, place them in the refrigerator overnight. Note that if the temperature indoors is warmer than the temperature outside, it may cause the buds on your cuttings to open more quickly.

Take students outside to find buds they wish to observe. (For the indoor version of this lesson, give each student a stem with a bud on it.) Before explaining details of the assignment, ask the students to share words that describe their buds. After getting several examples of adjectives, tell the students that they will write in journals about observing their buds. Give students copies of the handout.

Review topic sentences and how they set the tone for journal entries. Use the handout for the first page of the journal and staple additional pages to it. Alternatively, you can have the students use a folder or notebook for their bud journals. Encourage students to include illustrations of their buds and create graphs that represent how the buds changed. Wrap up the journal experience by displaying journals in the media center or on a bulletin board.

Assessment/Next Steps: Evaluate students on the correct use of parts of speech and reasonable observations of the buds. As enrichment to this activity, students can report bud openings on the Leaf Out section of the Journey North website (*www.learner.org/jnorth*).

My Friend, My Bud

Name: _____ **Date:** _____

In the wintertime, trees and shrubs that have lost their leaves form buds. Buds are protective shells from which leaves or flowers will emerge in spring. Depending on your location, these buds become noticeable in late winter or early spring.

Directions
Find a bud outdoors and become acquainted with it, just like a new friend! Tie a piece of colorful string around the branch to mark the bud you will watch. You will spend time with your new bud and write about it in journal entries.

Materials
notebook for science journal, pencil, magnifying lens, markers

Entry 1—Examine your bud carefully without touching it. Think about color, size, texture, shape, and any other features you can observe. Write a short paragraph in your journal: *My bud looks like . . .* Use five or more adjectives to describe your bud.

Note: It may take between a week and a month for your bud to open. If you look at a bud outdoors, the weather will play a major factor. Decide with your teacher if your Bud Journal will be daily, weekly, or three times per week. Use the ideas below to help you create entries in your journal. Remember to start each entry with a topic sentence, a sentence that tells a reader about the general idea of the entry.

Ideas for Journal Entries

Entry 2—Write about your bud's reaction to its new surroundings. Use two or three verbs. (For example, *When the wind blows, my bud bends!*)

Entry 3—Describe your bud's neighbors—nearby objects or animals. Use nouns for this entry and underline them. Be sure to include adjectives. (For example, *A small yellow and black <u>bird</u> landed on my <u>bud</u>.*)

Entry 4—Discuss the weather conditions in your bud's environment, underlining all prepositions. (For example, *My bud sways <u>in</u> the gentle breeze.*)

Entry 5—Describe your bud's changes since your first entry. Use figurative language such as similes and metaphors. (For example, *My bud grew as fast as my little brother!*)

Writing About the Seasons
Signs of Fall

Teaching Objectives: to practice creating a jot list; to make predictions; to order a sequence of events; to recognize seasonal changes in nature

Why/How to Use This Lesson: Writing about science helps students develop conceptual understanding of science topics such as the seasons (Klentschy 2008). As school begins, this lesson is a useful strategy for establishing science as the context for writing and note-taking throughout the year.

Topic: Seasons

Go to: *www.scilinks.org*

Code: OS020

Materials: handout, current year calendar, clipboards if available (optional: sidewalk chalk, patio calendar or place to draw a calendar on a sidewalk)

Procedures and Tips: As students begin to work on writing skills in science, they will benefit from writing about an occurrence that follows a natural, predictable sequence of events. This can help your students develop confidence in their skills of making predictions and writing about events in chronological order. This lesson—created for the beginning of the school year—has students writing about the progression of events in autumn.

Kinesthetic learners may benefit from the use of sidewalk chalk on a calendar patio that you have created, a sidewalk, or an unused driveway or parking lot. Moving around to make a large-scale calendar may be more effective than working with paper and pencil. If this is not available, put the handouts on clipboards and have students tour the school grounds. You can do the walkabouts several times during the course of a week or month to allow students to fill in their calendar. Because students are mainly jotting notes for this activity, it makes for excellent prewriting practice.

Assessment/Next Steps: Evaluate students' writing on reasonable observations and accurate portrayals of sequences of events. After this writing activity, consider the continuation of a science notebook that follows other observations, impressions, and inferences.

Writing About the Seasons
Signs of Fall

Name: _____ **Date:** _____

Directions

What happens to the world around you when summer ends and fall begins? Make a list of 10 changes in the world around you that fall could bring. Write down the date when you expect to see these changes. (Think of sights, sounds, smells, what animals do, and so on.) This is your prediction.

1. _The amount of daylight is less. Oct. 3_
2. _____
3. _____
4. _____
5. _____

6. _____
7. _____
8. _____
9. _____
10. _____

Use the calendar below to mark the actual date that each event happened on your list. Match the actual date to your prediction. If you have a calendar patio at your school, mark the events on the patio with chalk.

Sunday	Monday	Tuesday	Wednesday	Thursday	Friday	Saturday

The Diary of a Seed

Teaching Objective: to compare and contrast foreshadowing in literature with scientific predictions

Why/How to Use This Lesson: In a unit on plants or life cycles, this activity goes well with two lessons from chapter 4, "What Can You Learn From a Seed?" and "Germination Determination." As an integration strategy, this activity may be useful to elementary teachers searching for writing topics for their students.

Materials: handout, seeds, hand lenses, metric ruler, scale

Procedures and Tips: When it comes to getting your money's worth, seeds may be the best bargain a teacher can find! From harvesting seeds from dried flowers—such as marigolds, purple coneflowers, or sunflowers—to buying a dime package at the local dollar store, seeds may literally be a dime a dozen or better. In Chapter 4, the activity "What Can You Learn From a Seed?" offers seedling starter techniques for indoor use so students can base their writing on direct observations. Of course, for the adventurous teacher there is the option of gardening outdoors.

In this activity, students will learn to connect prediction in science to foreshadowing in literature as they work with 10 sentence starters. You may make up your own story starters to add to this activity.

Assessment/Next Steps: Students should compare their story of how a seed grows with what actually happens during this process. Evaluate students on reasonable portrayals of events. As a culminating activity, have students read their seed stories aloud or post them on a bulletin board or school website.

The Diary of a Seed

Name: _____ **Date:** _____

Directions

Think about what life would be like as a seed. Planting a seed can be an exciting adventure. To keep things interesting, plant one or two seeds without reading the package. Keep a journal about your seed for 10 days. The sentence starters on this page will help you get started each day.

Materials

seeds, potting soil, container, water, notebook for journal

Day-by-Day Sentence Starters for Seeds

Day 1: Today I was given to my new owner, _____, who thinks I am _____.

Day 2: After 24 hours, water has changed me by _____.

Day 3: Three adjectives that describe my new looks are _____, _____, and _____.

Day 4: I predict that I am a _____ seed because _____.

Day 5: I think I will have _____ leaves and _____ flowers.

Day 6: My roots are _____, and my stem is _____.

Day 7: As a plant, my purpose in life is to _____.

Day 8: Compared to my original seed, I am _____

Day 9: Growing seems so slow because _____.

Day 10: I knew water would help me grow, but I am surprised that _____.

Use these 10 sentence starters and add more if you continue writing in a journal for a long period of time. For example, you can compare your plant's growth to that of a human or wild animal. You can also look back—for instance, every fifth day you could compare what your plant is like to how it looked five days earlier. Be creative!

5

The Story of Life in a Tree

Teaching Objective: to identify story elements, such as characters in their setting and the logical sequence events of a plot

Why/How to Use This Lesson:

This exercise may be used as an introduction to writing fiction or a response to a story the class has read. It would also be a logical part of a unit on plants, seasons, or life cycles when paired with other activities, such as "My Friend, My Bud" or "Diary of a Seed." It can support writing instruction in middle school, when the science teacher coordinates with the language arts teacher, or integration for the elementary teacher.

Materials: handout, clipboards, bulletin board to post stories

Procedures and Tips: Divide the students into groups by having them call out the names of the seasons until everyone in the class has been identified as a winter, spring, summer, or fall. Have students form groups of four in which all four seasons are represented. Then

give each group the handout so that they can see the writing assignment.

Students can decide jointly on their characters, then each write separately about what would happen in the season they have been assigned. The students can then compare notes and work on story transitions from one season to another. They may wish to use notebook paper for a rough draft and then transfer the story to the handout.

For research, you might want to take students outdoors for observation of the current season, then talk about the other seasons so that all students can write their parts of the story at the same time. They can also explore information on the internet.

Assessment/Next Steps: Evaluate student writing for a logical sequence of events particular to the season assigned. Students may write this in a science notebook or journal and refer to it as the seasons change, comparing their story to the real changes of a tree in the school yard.

The Story of Life in a Tree

Name: _____ **Date:** _____

A tree is home to many animals and sometimes even other plants. Birds, squirrels, insects, and other animals live in trees, as do moss, lichens, and ferns. You are going to draw a large tree shape on paper and then write a story about what happens to all of the living things in a tree.

Directions

Choose a season. Now form a group with three other students who have chosen the other three seasons. Decide as a group which characters live in a specific tree. Fill in the blanks below, then write on the back of this paper about what the characters do during your chosen season. Work with your group to write the whole year's story on a large sheet of paper. First draw an outline of the chosen tree. Divide the tree into four equal sections. Label each section with the name of a season. Take turns writing your story on top of the tree shape.

Materials

large sheet of paper, water-based markers, pencil

Characters

Animals	Plants
_____	_____
_____	_____
_____	_____

Setting (my season): _____

The type of tree: _____

The part of the tree: _____

Events in the plot:

_____
_____
_____
_____
_____
_____
_____
_____
_____

Animal "Arti-Fact" or Fiction?

Teaching Objective: to write original fiction short stories using a writing prompt and an artifact from nature as story starters

Why/How to Use This Lesson: Using animal artifacts is an alternative for teachers who, for various reasons, cannot have living animals in the classroom. As students study characteristics of organisms (NRC 1996), they will benefit from hands-on experiences with artifacts such as feathers, exoskeletons, and other remnants that animals leave behind, particularly for students who do not have outdoor time in their home lives. This would be a useful activity in a unit on animals or life cycles.

Materials: handout, writing prompts developed by the teacher, feathers, snakeskin, turtle shell, leaves with caterpillar holes, other artifacts as desired, clipboard

Procedures and Tips: This is a strategy you can use indoors or outdoors, with or without artifacts. As the handout indicates, the ideal way to begin is to have students take a walk in the school yard to look for evidence of animal life. Come back to the classroom to debrief, and as a group make a list of what students have observed.

Ask students for ideas for a first line for a story that they could write about the animal that left one of the artifacts you discovered. This will model the opening part of the story-writing process for students.

Alternatively, students may use a starter of their own from their observations or one from the handout if nature has not inspired them; you could also post the starters from the handout on a bulletin board. Students can each choose the one of greatest interest to them and write it at the top of their paper. Students can then place the paper on a clipboard and go outside to continue writing while making direct observations.

Assessment/Next Steps: Evaluate students on reasonable inferences as they relate to the artifact or story starter that was used. To take this activity further, have a group of four or five students construct a food chain that includes the animals about which they have written. Once there are several food chains established, students may attempt to combine them into a food web.

Animal "Arti-Fact" or Fiction?

Name: _____ **Date:** _____

Directions

Go outdoors with your teacher. Look for animal artifacts on the school grounds. Some things you might find are an eggshell that fell from a bird's nest, skin shed by a snake or lizard, an anthill, rabbit droppings, deer tracks, leaves that have been chewed by a caterpillar, or the slimy trail of a slug or snail.

Choose one artifact and identify the animal that left it behind. Write a story about what happened to lead to the artifact being left there. You can also imagine what may have happened to the animal afterward as you write your story. Everyone has probably seen a movie or read a book that had animals in it. Think about those stories and what might really happen in nature to the animal and artifact you have chosen. The writing prompts below may help you get started. Use one of these prompts if it applies to an artifact you have seen.

Examples of Story Starters and Topic Sentences

- The slimy trail across the log was left after a young snail moved along slowly, as its shell got heavier.

- The soft, downy feather was the last baby feather shed by the gosling, which was officially a Canada goose.

- The jerking blue tail was not a sign that the skink had been killed but that its detachment defense had worked against the big snake.

- The empty cocoon still appeared peaceful, although no one knew if the luna moth had survived.

- The birds chirped happily in the trees. They enjoyed their meal from the now-empty feeder, which had been filled only yesterday.

- The squirrel lying in the parking lot had left behind three babies in her pine-tree nest. How would the babies survive without their mother?

The Novelization of Migration

Teaching Objectives: to practice the writing process; to use scientific facts in a work of fiction

Why/How to Use This Lesson: In an integrated unit on migration, this writing activity may fit well toward the end. Students can use facts they have learned throughout their study of migration. The complete study of animal regulation and behavior (NRC 1996) would include migration.

Materials: word-processing software or paper and pencil, books or internet access for research on migration, handout

Procedures and Tips: If students have followed the activities in the chapter in sequence, they will be ready to practice the writing process. They can also include science facts in their stories. The handout provides students with a writing assignment about migration and a checklist to help them refine their drafts.

Students may ask why they are writing about migration. Migration takes place over a period of time and has a sequence of events. Just as there is a sequence to the writing process, great writing often follows a sequence of events. Hopefully, there are future writers among your students, but in the case of those who need guidance, the scientific facts about an animal's migration will give them clear ideas for writing. For instance, a right whale spends the summer off the New England coast but comes to the coast of the Southeast in the winter to deliver its calf in warmer waters. This gives the student the setting for the story, as well as some basic plot elements to get him or her started. Guide students toward following the logical sequence of events found in the facts they learn, and help them apply the steps from prewriting to publishing.

Assessment/Next Steps: Evaluate students on their use of the writing process and the sequential use of facts regarding migration. Consider using the Journey North website (*www.learner.org/jnorth*) for related activities.

The Novelization of Migration

Name: _____ **Date:** _____

Directions

Why do animals migrate? Do different animals migrate for different reasons? Using the Journey North website (*www.learner.org/jnorth*), newspaper articles, magazines, or books, find out which animals migrate. Choose the animal that interests you most. Put yourself in your animal's place. Imagine that it is time for you to migrate, then write a fictional account of what happens to the animal on its migration. Base your story on scientific facts. Use the guidelines below to help you tell your animal's story. Your teacher will let you know how many paragraphs your story should be.

Writing Checklist

Prewriting

☐ Make a "jot list" of important facts about your animal and its migration. For example, find out its migration routes, life span, food sources, and more.

☐ Draw a story web or use another graphic organizer to put your thoughts in order.

Outline

☐ Write an outline that "maps" your story, just like your animal must map its migration.

Rough Draft

☐ Decide whether to tell your story in first person (*I, me, mine, we*) or third person (*he, she, they*). Remember to include scientific facts from your jot list in your story.

Proofreading

☐ Check for grammar and spelling. Spell-checking on a computer is not foolproof, so be sure to use a dictionary, too. Count the scientific facts in your story. How many did you include?

☐ Check to make sure your story has a beginning, a middle, a climax, and an ending.

Final Draft and Publishing

☐ After your story is corrected and polished, put it in final form.

☐ Draw at least one picture to illustrate your story.

Poetry in the Great Outdoors

Teaching Objectives: to recognize examples of acrostic poetry; to write original acrostic poems based on topics in nature

Why/How to Use This Lesson: Observation is an important skill for students to develop to participate in scientific inquiry effectively. Writing poetry about insects will give students the opportunity to practice observation skills and creative-writing techniques.

Materials: pencil, paper, handout, clipboards, insects in a container (optional)

Procedures and Tips: Read aloud from the handout the examples of acrostic poems using bugs and insects. Ask students to think about other insect names and nature-related words while you read. You may want to have some insects housed temporarily in a container for inspiration, or you can take your students on a school yard walk to look for insects. You can almost always find ants, even if you are surrounded by concrete, and grasshoppers and beetles are usually plentiful if your school has a grass field for recess.

Extensions: After the students have practiced acrostics with shorter words, such as *bugs* and *insects*, they should be ready to tackle longer words, such as *caterpillar* and *butterfly*.

You may want students to use their acrostic handouts from this activity as a reference when writing longer acrostics. Invite students to use *caterpillar*, *butterfly*, or another longer word of their choosing. As an added feature, require students to incorporate three to five facts into their poems. Students will have to do research to find these facts. Another option would be to provide a short list of science vocabulary words to use in each poem. For instance, for a poem on caterpillars, students could use words such as *larva*, *molting*, and *life cycle*. Vocabulary for the butterfly poem may include *chrysalis*, *pupa*, *nectar*, and *camouflage*.

Assessment/Next Steps: Make sure students use the science words in their poems in a way that demonstrates that they understand the definitions. When students are ready to publish the poems, you may wish to have them write the poems on paper that is cut in the shapes of their subjects.

Poetry in the Great Outdoors

Name: _____ Date: _____

Directions

One of the easiest types of poems to write is an acrostic poem, in which beginning letters on each line spell a word that is the subject of the poem. Look at the BUGS poem below, then change the words and write your own poem about bugs. Look around the school grounds and find out what bugs you in the insect world. Then write your poem!

Beetles are everywhere,   B_____

Under rocks and in the air,   U_____

Going here and going there.   G_____

Sometimes I just stop and stare. S_____

Poems do not have to rhyme; most of the time acrostic poems do not rhyme. Now that you have written one bug poem, it should not bug you to write another! This time, you can make it a little longer. Read the example for INSECTS. Then write your own poem in the space below.

I see insects all the time.
No, I am not scared of them.
So I try to go about my business,
Even when they get in my way,
Carrying my thoughts outside with them.
They must be happy in their carefree world,
So I get out of their way and let them be.

I _____
N _____
S _____
E _____
C _____
T _____
S _____

Listening to a Story About Wetland Ecosystems

Teaching Objectives: to identify common wetlands (ponds, rivers, streams, lakes, swamp, etc.) and the animals and plants they support; to have students listen closely to a story read aloud

Why/How to Use This Lesson: In a unit on ecosystems or biomes, using a book about wetlands would be useful as one of many resources from which students may gain a conceptual understanding of how living things interact in and around water. Reading a book aloud is often an effective opening for a unit of study and may be revisited at various times to frame student learning.

Topic: Wetlands
Go to: www.scilinks.org
Code: OS022

Materials: book on wetlands, handout

Procedures and Tips: A book such as *Salamander Rain: A Lake and Pond Journal*, by Kristin Joy Pratt-Serafini (Dawn Publications, 2001), helps students learn about the animals that live in ponds and other wetland areas. Read the story aloud and show the illustrations to students. Allow them to use the handout as you read to record their impressions and facts from the book.

Another way to use a book such as this one would be to put it in a reading center in the classroom and have students complete the handout after reading the book silently or in cooperative pairs.

Assessment/Next Steps: Evaluate students for reasonable answers to the handout questions about the story. If you have an outdoor classroom or another suitable area on the school grounds, think about creating a bog garden that is home to plants that love moisture (pitcher plants, Venus flytrap, etc.) after you finish this assignment.

Listening to a Story About Wetland Ecosystems

Name: _____ **Date:** _____

Directions

Listen carefully as your teacher reads from a book about a wetland ecosystem. Record what you hear and think about what you might see (animals and plants). Make notes on this page.

1. What are some sounds you might hear in this wetland area?

   _____

   _____

2. What are some different ways animals in this wetland move?

   _____

   _____

3. What are the plants like in the area? Describe one or two.

   _____

   _____

4. Think of a place like the one in the story that you have seen yourself. How are these places alike? How are they different?

   _____

   _____

5. Choose one of these activities: (a) Sketch a scene from the book, or (b) Draw a bird's-eye view of the wetland (like a map) and label the different areas. Use the space below.

5

Learning About Caterpillars and Butterflies

Teaching Objectives: to identify and order the stages of life for a butterfly; to have students listen closely to a story

Why/How to Use This Lesson: In an integrated unit on life cycles, a book about caterpillars and butterflies serves as one of many sources for students. Reading a book aloud is often an effective opening for a unit of study and may be revisited at various times to frame student learning.

Materials: book, handout, caterpillars to raise in the classroom (optional)

Procedures and Tips: Identify students' prior knowledge by having them tell you or write down what they know about the life cycle of a butterfly. Read the book to the class or have students read multiple copies or even various titles.

Students may complete the handout while listening to or reading the book, or they can work on the handout afterward to strengthen their memorization and listening skills.

Consider having live caterpillars in the classroom while you read about the life cycle of butterflies. Students can compare and contrast the events in the book with those with the classroom caterpillars.

Assessment/Next Steps: Evaluate students for reasonable answers to questions on the handout. If your school has a butterfly garden, take the students outdoors to identify plants that attract butterflies and caterpillars. If you do not have an outdoor classroom established, identify plants from the book and think about planting some in a flowerpot to place outside near your classroom windows.

Learning About Caterpillars and Butterflies

Name: _____ **Date:** _____

Directions

Look carefully at a book about caterpillars and butterflies. Think about how these insects change. Then fill in the blanks.

1. What does the caterpillar look like? Write three words to describe it.

   _____   _____   _____

2. How does the caterpillar change during its life? Write a complete sentence that tells about the change.

   _____

   _____

3. Caterpillars eat leaves, and butterflies drink nectar from flowers. How do these actions help them stay alive?

   _____

   _____

4. Draw the stages of the butterfly life cycle. Label the stages: egg, caterpillar, chrysalis, and adult. Include arrows if that will help you.

   1

   4

   2

   3

5

Solving an Ecological Mystery

Teaching Objectives: to identify cause and effect in literature and nature

Why/How to Use This Lesson: In an integrated study of how man interacts with the environment, reading a book about an ecological mystery benefits students by sharpening their ability to identify cause-and-effect relationships. Consider reading this type of book to students in the weeks leading up to Earth Day as preparation for working on local ecological issues.

SCiLINKS
THE WORLD'S A CLICK AWAY

Topic: Solving Environmental Problems

Go to: *www.scilinks.org*

Code: OS023

Materials: book, handout, newspapers

Procedures and Tips: This lesson is inspired by the book *Who Really Killed Cock Robin?* by Jean Craighead George (Simon & Schuster, 1973) but may be used with any book that presents an ecological mystery. Have students read chapters of the book in cooperative pairs or individually for homework. You may also choose to read the book out loud over a period of time. Students should have the handout available to them before you begin reading the book so that they can record answers to the questions as you read.

Students may participate in class discussions or even a debate about the possible causes of ecological problems. You may also want to have your students use newspapers to identify local problems in the environment that could lead to situations such as those in the book they read. You may wish to have your students list their predictions for the conclusion of the story on chart paper; then the class can refer to the chart after reading the book to check for the accuracy of their predictions. Use this discussion as a springboard to talk about the reasons behind what has taken place in the plot.

Assessment/Next Steps: Evaluate students on reasonable answers to questions on the student handout and their ability to support reasoning with evidence from the story being read. After using a book to spark discussion on ecological problems, have students research the local environment and create a service-learning project in which students can make community improvements.

Solving an Ecological Mystery

Name: _____ **Date:** _____

Directions

Listen for clues in the mystery story that your teacher has chosen, then think of the mysteries in the plot. What do you think will happen next?

1. List five important pieces of background information.

   1. _____

   2. _____

   3. _____

   4. _____

   5. _____

2. What is the main problem or mystery in the story?

   _____

   _____

3. Who are the main characters? Are they good guys, bad guys, suspects, victims, or innocent bystanders? Write each character's name and category. Add a second category if you need more than one way to describe the character.

Solving an Ecological Mystery

Character	Category
(example: Frog)	(example: Victim)

4. Listen to enough of the story to guess the solution. Then draw a map, flow chart, or story web to show what you think happened and how the mystery will be solved.

5. Listen to or read the rest of the story. Did you guess correctly?

5

A Migration Story

Teaching Objectives: to identify the sequence of events in a story and the sequence of events in migration; to have students listen closely to a story

Why/How to Use This Lesson: In any study of animal regulation and behavior (NRC 1996), students can benefit from hearing stories of migration. Consider sharing the story in increments so that students have time to make predictions and examine the sequence of events closely. For instance, read aloud for about 10 minutes each day, or if using a video, show about 20 minutes every third day over a couple of weeks. This would fit well with the migration lessons from the previous two chapters.

Materials: book, handout, video (optional), internet access (optional)

Procedures and Tips: Inspired by the book and movie *Fly Away Home* by Patricia Hermes (Newmarket Press, 2005), this lesson helps students identify the sequence of events in nature and literature. In birds' lives, imprinting, learning to fly, and following parents on journeys are often the events in the migration sequence. Students may use the internet to compare and contrast the migration of other animals with one another.

Give students the handout before you begin reading your selected book. If you choose to read the book version of *Fly Away Home* to your students, you may decide to watch the video (or some video excerpts) of the movie after you have finished reading.

Make sure to preview the movie and determine if it meets your school system's guidelines. Students may even decide to research the real girl behind the story in the movie.

Assessment/Next Steps: Students should take notes from the book or movie and compare the sequence of events in literature to those necessary for successful migration in nature. Evaluate students on their reasoning as they answer questions comparing the story to actual migration.

A Migration Story

Name: _____ **Date:** _____

Directions

Think about the scientific facts about the animal(s) in the story that your teacher has chosen. Then answer these questions as you learn more about migration.

1. Why does the animal in the story migrate?

_____

_____

2. What are the advantages and disadvantages of migration?

Advantages	Disadvantages
(example: warmer weather)	(example: long-distance travel)

3. Draw a map below that shows the route you think the animals will take. Use a dotted line in pencil for your prediction. As you listen to more of the story, use a red crayon or pen to show what route the animals in the story actually took.

4. Research this topic: How does this animal's migration compare to other animals' migrations? What do the animals have in common? What are their differences? Write your report on another sheet of paper.

5

Reading From Nature Journals

Teaching Objectives: to identify ways in which setting can inspire the written word; to keep a nature journal for several days; to have students listen closely to a story

Why/How to Use This Lesson: Writing is an integral skill to scientists, and students can benefit from reading examples of science journals that record observations about animals and plants from notable naturalists such as John Muir. Consider introducing science writing and journaling to your students by reading some examples aloud. If you are going to have students keep a nature and science journal throughout the year, start the year with this lesson, and revisit journals of the "masters" often to remind students how to record detailed descriptions.

Materials: book, student journals

Procedures and Tips: Collect some of the writing of John Muir, one of the foremost authors of nature journals. Read some selections aloud to your class, or have them read selections on their own. After completing several of the assignments in this chapter, students should be ready to keep a nature journal, complete with descriptive language and sketches. If possible, take students outdoors regularly for a period of several days or assign outdoor observation as homework. If it is not possible to go outdoors regularly as students work on their journals, have students observe out a window, use a photograph of an outdoor scene, bring in some natural objects (rocks, branches, feathers, flowers), or simply imagine what they would see outdoors. Ask students to read selections from their own journals aloud to partners, and ask the partners to illustrate the descriptions of the outdoors.

Assessment/Next Steps: Evaluate students on their detailed descriptions of natural events or objects from nature. If you have students keep a journal as part of your regular school-year assignments, consider giving them the option of making it a nature journal based on this activity, and have them keep the journal for the rest of the school year. If your class has a website, students may take turns blogging in a group nature journal.

NATIONAL SCIENCE TEACHERS ASSOCIATION

Social Studies

Humans and the Outdoors

The history of humans interacting with the land is the perfect context for a number of outdoor learning experiences. In fact, for many children throughout history, the outdoors was a classroom—although maybe not officially. Think about how humans have relied on the Sun to tell time or on plants to provide food sources, shelter, and clothing. As students map the school yard, discuss with them how animals "map" the outdoors. Compare animals' migration routes with humans' travel routes. Use your state standards and the teacher's edition of a social studies textbook to get ideas for relating history, geography, and social studies to the outdoors. Here are a few ideas to get you started.

6

Planting a Native American Garden

Teaching Objective: to explore Native American culture through gardening

Why/How to Use This Lesson: Science as a human endeavor and the history of science (NRC 1996) are important for a student's ability to relate science and social studies. Native Americans practiced planting techniques that used available resources but did not exhaust the land, what we might refer to today as green. Use this lesson to integrate science into a social studies unit on Native Americans or an integrated unit on Thanksgiving.

SCI LINKS.
THE WORLD'S A CLICK AWAY
Topic: Plants as Food
Go to: *www.scilinks.org*
Code: OS024

Materials: seeds for corn, squash, and beans; bamboo poles or long sticks; garden plot or large container, handout

Procedures and Tips: Students plant beans, squash, and corn in a mound of soil or large garden container, creating a "three sisters" garden. Students can mimic a teepee by leaning three poles in together, allowing the beans and squash vines to travel up the poles. Do not let corn plants end up "under" the teepee. You may wish to start plants indoors and transfer seedlings outdoors to either a garden or a large container. Provide an additional selection of seeds so that students can compare two different sets of seeds to determine whether other plants benefit from the same planting methods. Facilitate a way for students to share their findings, either electronically or on a bulletin board.

Assessment/Next Steps: Evaluate student reasoning based on their predictions from the student handout and the ability to explain the results of their gardening experiences. Further integrate this lesson with writing by having students write a how-to manual for planting with Native American gardening techniques.

Planting a Native American Garden

Name: _____ **Date:** _____

Before the modern era of farming, many methods were used to maximize the nutrients in the soil and avoid repeatedly planting the same crop in the same place.

Do Some Research

Native Americans were said to have planted "Gardens of the Three _____."

The three types of vegetable seeds they planted in these gardens were

_____, _____, and _____.

Why did they plant the three types of seeds in the same mound?

_____

_____

Make Some Predictions

Would a Native American garden planted now have the same benefits as it did historically? Why or why not?

_____

_____

What other plants might benefit from being planted together? How?

_____

_____

Experiment

1. Plant a "three sisters" combination of plants. Record your observations on one side of a two-column journal and include either sketches or digital photographs.

2. Plant another similar combination of plant seeds that you choose from those available. Keep the same type of records for these plants in the second column of the journal beside the three sisters information.

3. Compare the results and share them. If you have digital photographs, share them in a multimedia presentation or post to a class website. If you have sketches, scan them if possible, or use them to create a bulletin board.

It's About Time and Human Sundial

Teaching Objectives: to use the Sun to measure time; to practice measuring distance

Why/How to Use These Lessons:

This lesson fits into a variety of integrated units, from discussions of objects in the sky to how humans lived before modern inventions. Combine this lesson with the previous lesson on the three sisters garden in a unit about Native Americans.

Scientists Who Use These Skills:

anthropologists, naturalists

Materials: paper plates, drinking straws, tape, pencils, patterns, compass, handouts

Procedures and Tips: First, have students research the history of telling time. Next, give students the handout on page 117 and ask them to answer the background questions and write a hypothesis about how to make a sundial. Give students the necessary materials and let them make sundials (see handouts on pp. 117 and 118).

Take students outside on a sunny day to test their creations. Check the sundials with a watch. You will need to get outdoors on the hour for multiple hours, so consider sharing this activity with another teacher. If you teach only science or math, it should be fairly easy to take groups to mark different hours.

When marking shadows, students use the straw as a *gnomon* (shadow caster). They will need to lean the straw slightly toward north (the angle of the gnomon depends on your latitude). If you do not have a compass, look on a local map or use a car with an electronic compass to find true north. After your students have made a shadow on the plate with the gnomon, use your watch to help them indicate the time with an hour line.

If you want to extend this experience, make a human sundial (p. 119) using this procedure:

1. Explain how to make a human sundial. For example, say, "You can make a human sundial. A gnomon is the part of a sundial that casts the shadow. Instead of using a straw as a gnomon, we will use a person."

2. Have students make a mark on the ground in a sunny space. You may want to use a brick or concrete block so the mark does not move easily. If you want to use a paved area, you can mark your spot with chalk.

3. Find the direction north. Draw a line straight out from the student (gnomon) toward north as the student stands on the spot you marked. This line will be noon. Try to go out at noon to test the mark.

4. Make other marks by having students stand on the spot every hour on the hour. It may take you several days to complete this task.

Assessment/Next Steps: Evaluate students' reasoning on the Think About It questions on the handout (p. 117) and their abilities to adjust the sundial designs to measure time accurately. With help from a carpenter or hardware store employee, help students design sundials from more permanent materials for the outdoor classroom or to take home.

It's About Time! Sundial Activity

Name: _____ **Date:** _____

Materials
paper plate, drinking straw, tape, pencil, pattern (see p. 68), compass

Directions
Gather your materials and complete the steps below.

1. Write a hypothesis that proposes how you could make a sundial.

   If _____,

   then _____.

2. Using the materials listed, make a paper plate sundial. Your sundial can be constructed any way you want, as long as you believe it will work.

3. Test your sundial in a sunny place outdoors. Place the dial facing north and use the straw to cast a shadow. Mark the hour on the face of your sundial in pencil. You can test it again at the same time during two other days to check its accuracy.

4. Draw a picture of your sundial at work.

5. Mark the other times on the sundial. What percentage of accuracy do you think your sundial has?

Think About It!
1. How do people today determine the time when they do not have a watch or clock?

2. Before the clock was invented, how did people measure time?

3. How could you make an accurate sundial if you were in the wilderness?

6

Sundial Pattern

Name: _____ **Date:** _____

Directions

Use this pattern to lay out your sundial on a paper plate. Cut out the pieces. Glue them in place after testing your sundial and comparing your marks to the time on a watch. The shadow cast by the sun is specific to your location. Take the sundial outside to check it every hour on the hour. You may have to spend a few minutes doing this each day for a week.

North

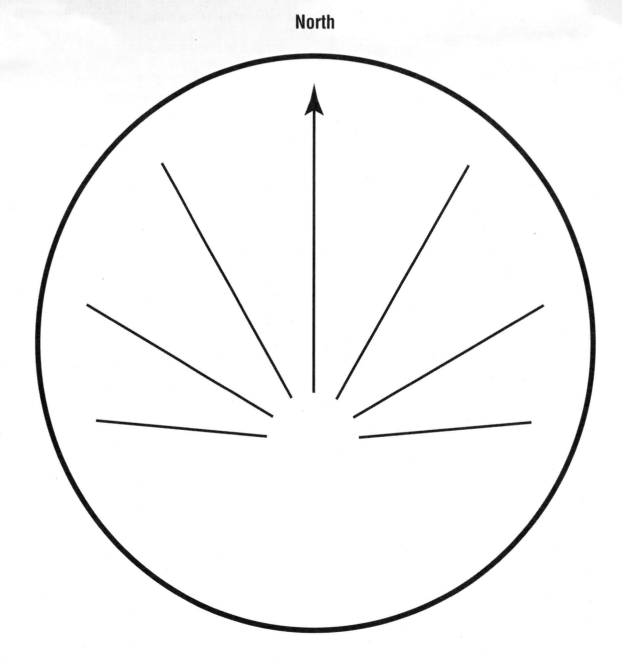

Make a Human Sundial

Name: _____ **Date:** _____

Directions

Work with your class and follow your teacher's instructions to make a human sundial. Test the human sundial against the paper plate sundial. Compare the results and answer the questions below.

1. At noon, in what direction is your shadow cast?

   _____
   _____

2. What is the difference in minutes in the accuracy of the two sundials?

   _____
   _____

3. Which sundial is more accurate? Why?

   _____
   _____

4. In what ways are the shadows of the human sundial the same or different during different times of the day?

   _____
   _____

5. Are the angles between each hour the same? Explain.

   _____
   _____

6. How is a sundial the same as a traditional clock and how is it different?

   _____
   _____

7. Make a third sundial by sticking a twig in the ground. Let the twig cast a shadow. Could you do this if you were in the wilderness and needed to know the time? Describe what you would do in that instance. What materials would you need if the sky was overcast?

   _____
   _____
   _____
   _____
   _____
   _____
   _____
   _____

Planting Historical Herbs

Teaching Objective: to teach students about historical uses of herbs by planting and harvesting various herb plants

Why/How to Use This Lesson: In a unit on colonial America, integrate science by helping students discover how herbs were used for medicinal purposes and cooking. Compare and contrast with modern uses of herbs. Some herbs also serve as host plants for butterflies and could be planted as part of a unit on butterflies or life cycles.

Topic: Medicines From Plants

Go to: *www.scilinks.org*

Code: OS025

Materials: herb seeds or plants, garden area or planting containers, hand clippers or scissors, string or twine, handout, books and internet access for research

Procedures and Tips: Help students discover how herbs were used in the past. Many herbs were used for cooking, medicine, or cleaning and dyes. The handout for this activity helps students discover the historical uses of herbs. This lesson should be used as an accompanying activity to planting an herb garden. Purchase small herb plants at a local gardening center or grow them using seeds. Seeds may be started indoors and moved outdoors later. In some cases, it may work to have potted herbs in the classroom near a window. There are many commercial products that serve as kitchen planting containers for herbs that would also work for classroom use. Follow the accompanying planting instructions or instructions on seed packets.

Assessment/Next Steps: Evaluate students' responses to prompts on the handout. Students should be able to verbalize the various ways to use herbs. You may wish to have students work with the school cafeteria manager or a local chef to learn how to use herbs to prepare a meal or contribute herbs they have grown to a community food bank.

Planting Historical Herbs

Name: _____ **Date:** _____

Have you ever heard of rosemary or dill? Do you know how they are used? Why is a dill pickle called a dill pickle? Many herbs are used in cooking and cleaning. In the past, many herbs were grown in home gardens and used in daily living.

Use a book or other resource to find the names of at least five herbs. List them here.

1. _____
2. _____
3. _____
4. _____
5. _____

Will these herbs grow where you live? What is their growing range, and what weather extremes can they tolerate?

_____
_____

Research the history of two of the herbs you have listed. How were they used historically? Were they used for medicine, cooking, or household products such as soap or dye? Write a brief description of their uses.

_____
_____

Choose at least one of the herbs you have studied as the subject for a historical journal or diary entry. You may need to use the back of this sheet or another piece of paper for this. Choose a date (including the month, day, and year), and write about how you—as a fictional, historical character—use the herb. Underline the name of the herb in your story. **(Example: September 1, 1860—I am a young mother who lives on farm. I have a new baby son who needs clothes. I am going to use the <u>indigo</u> we grow to dye the baby clothes blue.)** Write at least one paragraph and include a sketch of how the herb was used and what the herb plant looks like. Stay true to the era that you describe.

_____
_____
_____
_____
_____
_____
_____

Be an Archaeologist!
Make a Grid

Teaching Objectives: to practice making use of a scaled drawing; to use a grid to measure the distance between objects in a mock dig site

Why/How to Use This Lesson: Integrate math, science, and social studies in a unit on archaeology. This lesson helps students develop an understanding of how the disciplines intersect and why scientists and even social scientists cannot adequately do their jobs without math. You may want to have students draw their grids indoors and then create a larger grid outdoors.

Scientists Who Use These Skills: archaeologists, paleontologists

Materials: handout, metric ruler, pencil; Building mock dig site—meterstick, yardstick, or tape measure; play sand from local home improvement store; landscape timbers; nails; string; landscape plastic; heavy-duty stapler; clay "artifacts"

Procedures and Tips: Guide students in a discussion about why scientists working on an archaeological dig need math skills. You may wish to have students research a current archaeological dig on the internet so that they can visualize the techniques. Sometimes you can find a site with real-time video streaming. After you have discussed the math needs of archaeologists, give students the handout and have them work on the exercises.

Assessment/Next Steps: Evaluate students on the accuracy of their scaled drawings and their reasoning for why scientists need this skill. Extend the lesson in a long-term project by having students help you build a mock dig site in which they can practice using a grid. Read more about this activity in Chapter 1.

Follow this procedure for the activity:

1. Determine the size of the area using the meterstick, yard stick, or tape measure.

2. Build an edge around the perimeter with landscape timbers.

3. Line the area with landscape-grade plastic, and staple the plastic to the interior edge of the timbers.

4. Use a nail to poke random holes in the plastic to allow potential rainwater to drain.

5. Fill the plastic with sand to a depth of about 30 cm, or use enough sand to cover potential "artifacts" you would like to hide.

6. Buy plastic bone sets from a dollar store or science company catalogue. Buy clay pots to hide as well. The bones may be separated and buried, and the pots may be broken and buried.

7. Have students use small trowels, brushes, and other tools to remove enough sand to determine the location of the "artifacts."

8. Put nails in the landscape timbers in even distances apart and create a string grid over the dig so that students can sketch a scaled drawing of where each item has been found.

9. Items may be retrieved, and students may work in groups to rebuild skeletons or pots.

Be an Archaeologist!
Make a Grid

Name: _____ **Date:** _____

Directions

When an archaeologist digs up artifacts, the scientist must carefully make a record of where the artifacts were found. This helps each scientist accurately reconstruct the past. See if you can make mathematical sense of this scientific discovery. The space between the lines represents an actual space of 10 cm. Draw the strings that would go horizontally and vertically over the dig. The grid over the dig site represents one made of string and stakes. Using the key, answer the questions below.

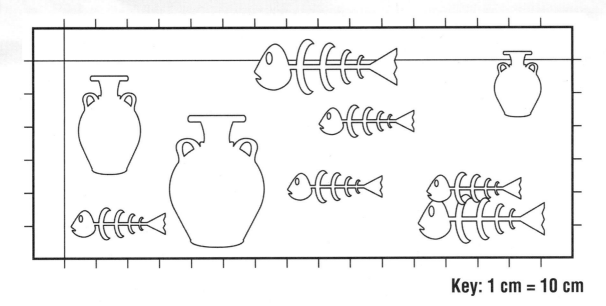

Key: 1 cm = 10 cm

1. What is the average length, in centimeters, of the artifacts?

   _____

2. What is the height and width of each pot?

   _____

3. What are some possible explanations for why the artifacts and pots ended up in this pattern? Pretend you are the scientist who made this discovery. Write a journal entry and tell the events of the past that led to the layout of this find. Write your entry on the back of this page.

More Digging!

Working with a team of three or four students, set up an archaeological dig in your outdoor classroom area for other students to investigate.

6

Rocks and Man

Teaching Objective: to observe the properties of rocks and how they have been used by humans in various ways (tools, weapons, building materials, etc.)

Why/How to Use This Lesson: Integrating science and social studies makes this lesson ideal for elementary teachers with science curriculum objectives on the properties of rocks and social studies objectives on man's use of natural objects as early tools. Middle school science and social studies teachers could cooperate and teach lessons on properties of rocks in science and rocks as tools in social studies.

Topic: Rocks and Human History

Go to: *www.scilinks.org*

Code: OS027

Materials: rocks, objects made from rocks or photographs of same, hand lens, field guide to rocks (or internet access to websites about rocks), handout

Procedures and Tips: If possible, expose your students to real historical objects made from rocks. Check with a local university, museum, science center, or educational agency to find loaner archaeology kits that include historical artifacts made from rocks. A Native American center might also have artifacts or replicas that you could borrow or photograph. Many people collect arrowheads, spear tips, or tools made from rocks. Invite a collector to visit your class and share his or her collection. If it is not possible to have the physical objects, use photographs from a book or the internet to share with students. Take students on a walk through the school yard to look for rocks. As students find rocks, initiate a discussion on what use humans might have for that particular rock. Discuss the use of rocks as weapons, tools, and building materials. A unique historical use of some larger rocks was as a millstone, which was part of a mill that ground corn into cornmeal or wheat into flour. Use the student handout for students to record their thoughts and sketch the use of rocks.

Assessment/Next Steps: Evaluate students' responses on the handouts for reasonable answers. Once students have recorded and sketched one of the uses of rocks they discovered, they are instructed to compare and contrast with another student. You may wish to have students work in pairs to create historical reproductions or models of tools that were made with rocks. If possible, involve a historian from a local museum or university.

Rocks and Man

Name: _____ **Date:** _____

From the beginning of human civilization, rocks have been important to the relationship between man and the outdoors. Can you find a rock that may be used for another purpose? Does anyone in your family collect rocks or items made from rocks? Think about other ways in which you have seen people use rocks.

Write your thoughts or observations about the following uses of rocks. Give specific examples if you know of any, or look up answers in a book or on the internet.

Tools _____

Weapons _____

Fire starter _____

Building materials _____

Use this space to sketch one of the uses of rocks above or another one that you have thought about or discovered in research. If you have any examples of rocks for other purposes, try to sketch what the object looks like. If possible, determine the type of rock using a field guide, book, or the internet. Compare and contrast with a classmate.

6

This Land Is Your Land, This Land Is My Land
How Humans Use Fences

Teaching Objectives: to explore how building fences and walls creates perceived borders in natural spaces; to build a fence or model of a fence

Why/How to Use This Lesson: As with the other lessons in this chapter, this lesson is designed to help teachers integrate social studies with science. This lesson would fit well into the study of cultures that have a significant historical wall. Historic sites that might fit into this type of instruction include the Great Wall of China, the Tower of London, or Wall Street in New York City. A unit that ties human activity to the environment may include this lesson as well as "Rocks and Man" and "Solving an Ecological Mystery."

Materials: wood, rocks, bricks, or other fence-building materials (or smaller representative materials for models); internet or books for research; handout; digital cameras (optional)

Procedures and Tips: Help students explore the historical significance of fences and walls by researching photographs in books or on the internet. If possible, students may use digital cameras or cameras on cell phones to photograph fences at their homes, in their neighborhoods, and in surrounding areas. If there are any fences on school grounds or visible from school grounds, take a walk to look at these fences with your students. Discuss the purpose of each fence. Is it there to keep something or someone in or out? Some fences might be designed to keep students on a playground, intruders from coming onto school grounds, or students away from air conditioners or other equipment. Compare and contrast local fences with those with historical significance (such as the Great Wall of China, the Berlin Wall, the Dutch wall that Wall Street was named for in New York, or walls around historic forts). If you use the "Rocks and Man" activity, tie the two together by discussing rock walls and fences. If you use the mapping activity, discuss how fences are often placed along borders that show up on political maps. If you are building an outdoor classroom or have an existing outdoor classroom that needs a fence, tie that project to this lesson. If not, have students create a model of a fence or wall, such as an early American split-rail fence or a Victorian picket fence.

Assessment/Next Steps: Evaluate students on reasonable responses to answers on the student handout. After completing this assignment, students may also create virtual models with computer programs or scale models. Have students share their models with the class or display them in the school.

This Land Is Your Land, This Land Is My Land
How Humans Use Fences

Name: _____ **Date:** _____

Fences and walls have been used by humans for thousands of years. Name several reasons why humans have built fences and walls.

Name one historical wall or fence of significance. Why is/was it important? What is/was its purpose?

_____

_____

_____

What would happen if there were no fences? What would be the consequences of removing all the fences in your community or city?

_____

_____

_____

What materials are used to build fences today, and how has that changed throughout history?

_____

_____

_____

Sketch or photograph a fence in your community. Use the back of this sheet if necessary. What is the purpose of the fence you drew or photographed?

Create a Map!

Teaching Objective: to create a map of a familiar area to practice map skills

Why/How to Use This Lesson: Geography skills are used by scientists, so there is a natural connection for mapping and scientific endeavors. Middle school science teachers can support social studies teachers by having students complete this activity while the social studies teachers cover map skills. Elementary teachers who have curriculum objectives on maps can give the students a practical application of map skills by having them create maps of outdoor classrooms and the school yard.

Materials: handout, pencil, ruler, a variety of maps, digital cameras (optional), computer access for map software or websites (optional)

Procedures and Tips: If you have an outdoor classroom, ask students to create a map of the outdoor classroom. If you do not have an outdoor classroom, students may create a map of the school grounds, their neighborhood, or a local park. Students may need metersticks or tape measures to establish a scale for the objects on their map. Students may create a small map on a single sheet of paper or use newsprint or butcher paper to make a larger map that folds up. Share some maps with your students to give them ideas. Get a free road map from a visitors center or transportation department. You may also want to get out students' social studies textbooks, atlases, and other printed maps. Students should study features of various maps to determine how symbols, scales, keys, and other map features help people find their way.

Assessment/Next Steps: Evaluate students' maps for reasonable representations and scales based on the areas they map. Take it further by incorporating satellite images of the area or studying existing maps. Detailed local maps often are available from local real estate agents. The GLOBE (*www.globe.gov*) curriculum activities also include mapping and additional teaching ideas.

Create a Map!

Name: _____ **Date:** _____

A map is simply a guide to an area using representations of real objects to create a one-dimensional model on paper or a computer screen. Create a map of an area as assigned by your teacher or of your choice. If you have an outdoor classroom, you may wish to map that. If not, your school yard, street, or neighborhood might be good choices. You could even create a map of your school building. Use a ruler, meterstick, or tape measure to create a scale for your map. You may use the back of this sheet or a larger piece of paper if necessary.

Make a list of the steps you will take to create your map. Think of all the resources you may use. If you have a paper map or internet access, those may help you get started. You may have more than five steps if they are needed.

1. _____

2. _____

3. _____

4. _____

5. _____

List the materials you will need:

_____
_____
_____
_____
_____
_____
_____
_____
_____
_____

Migration Mapping

Teaching Objectives: to identify animals' migration routes; to compare and contrast two animals' migration routes

Why/How to Use This Lesson: In an integrated unit on migration or animal regulation and behavior (NRC 1996), integrate social studies and map skills by having students use a map to show the migration routes of various animals. Pair this with the next lesson, "Paper 'Migration' of Monarch Butterflies" and track the migration on the internet.

Materials: handout, ruler, blue and red markers, research materials (books or internet access to *www.learner.org/jnorth*)

Procedures and Tips: Have students observe in the school yard and then research migration routes on the Journey North web-site or using other sources. Ask students to use the map on the handout to trace the migration routes of any two animals that migrate through your area. If there are numerous animals with migration routes through your area, you may wish to divide the students into groups and assign two animals to each group. After students complete the activity, they can post their maps in the room or put their lines of migration onto a group map. For the group map, use a copy machine to enlarge the map on the reproducible, and use additional colors to make it easier to see different routes.

Assessment/Next Steps: Evaluate students on their accurate representations of two migration routes on their maps. Students may communicate with other students around the world to track the migration of various animals through the Journey North website (*www. learner.org/jnorth*).

Name: _____ **Date:** _____

Directions

Identify two types of animals that migrate through your state. Use the map below to trace the migration routes of these animals. Highlight the location of your home in yellow. Put an asterisk to show the location of your town or city. Make dashes in red for one migrating animal. Make dashes in blue for the other animal. Illustrate your map with small pictures of your animals along their routes. How long does it take each animal to complete the one-way journey? If they left your school at noon today, where would they be at noon tomorrow?

Animal traveling on blue migration route: _____

Possible location of this animal in 24 hours: _____

Animal traveling on red migration route: _____

Possible location of this animal in 24 hours: _____

Think About It!

The Migrating Monarch Butterfly

Every year, millions of monarch butterflies travel from the United States and Canada to the mountains of central Mexico. Would you like to take part in this amazing trip? Plant some milkweed seeds in your school yard! The butterflies find everything they need from this one plant.

It takes several generations of butterflies to travel the 2,000 miles to Mexico. Find out more about this migration on the Monarch website (*www.monarchwatch. org*), then trace the migration routes of these insects on the map above.

Paper "Migration" of Monarch Butterflies

Teaching Objectives: to link cultural and geographical effects of migration to science content; to practice map skills

Why/How to Use This Lesson: This lesson fits into an integrated unit on migration or for the teacher who has social studies curriculum objectives on the Americas or Mexico. A fall unit on butterflies would also be a good fit for this activity. You can use this lesson to support instruction on animal regulation and behavior (NRC 1996).

Materials: handout, crayons or markers, large mailing envelope and postage, internet access (*www.learner.org/jnorth*)

Procedures and Tips: This project usually has a mid-October deadline, so plan ahead by visiting the Journey North website during the summer or at the start of the school year to locate all of the information on this paper-butterfly exchange between American and Mexican students.

When creating their paper butterflies for the "migration," students may use the pattern provided on the reproducible to make their paper butterflies, or they can create their own monarch butterfly designs. The finished work must be flat, not multidimensional.

Students should include a friendly message to Mexican students in which they might encourage their new friends to preserve the habitats of the monarch butterfly and other wildlife.

When all of your students have colored their butterflies and written their messages, gather the butterflies and mail them in a large envelope to Journey North. Make sure to include a self-addressed, stamped envelope of about the same size as your mailing envelope. This will carry Mexican butterflies from the Journey North offices back to your school in the spring. Journey North orchestrates this activity for teachers.

The "migration" is supported by the internet site, which provides updates on the locations of the paper butterflies, making comparisons with the routes and progress of the real butterflies. In the spring your class will receive paper butterflies from Mexico and perhaps other places in the United States or Canada. This will allow your students to create interesting maps that show where the butterflies originated and the routes they took to get to your school.

Assessment/Next Steps: Depending on the instructions given to students, evaluate their ability to create reasonable representations of monarch butterflies and follow directions for the accompanying note. To follow up, use the Journey North website and consider using a pull-down or wall map to chart the locations of the paper butterflies.

Paper "Migration" of Monarch Butterflies

Name: _____ **Date:** _____

Each year, students across America and Mexico send paper butterflies to each other. This "migration" is in honor of the real monarch butterfly migration.

Directions

1. Color or paint the butterfly on this page. Show correct monarch colors.

2. Write a message to the Mexican student who will get your butterfly. Ask the student to help protect the monarch butterflies and their habitat.

3. Look at the Journey North website (*www.learner.org/jnorth*). Follow the directions to send in your paper butterflies. If you live in the United States or Canada, you will do this step in the fall.

4. Paper butterflies will "migrate" back to your school in the spring. They will be sent from students in Mexico. The paper butterflies will arrive at the same time the real butterflies start to fly north!

References

Keeley, P., F. Eberle, and C. Dorsey. 2008. *Uncovering student ideas in science: Another 25 formative assessment probes.* Vol. 3. Arlington, VA: NSTA Press.

Keeley, P., F. Eberle, and J. Tugel. 2007. *Uncovering student ideas in science: 25 more assessment probes.* Vol. 2. Arlington, VA: NSTA Press.

Klentschy, M. 2008. *Using science notebooks in elementary classrooms.* Arlington, VA: NSTA Press.

Louv, R. 2005. *Last child in the woods.* Chapel Hill, NC: Algonquin Books.

National Council of Teachers of Mathematics (NCTM). 2000. *Principles and standards for school mathematics.* Reston, VA: NCTM.

National Research Council (NRC). 1996. *National science education standards.* Washington, DC: National Academies Press.

National Science Foundation (NSF). 2000. *Environmental science and engineering for the 21st century.* Arlington, VA: National Science Foundation.

About the Author

As a science teacher in elementary and middle schools, Steve Rich created two outdoor classrooms that were honored with NSTA awards—the Ciba Exemplary Science Teaching Award and the Ohaus Award for Innovations in Science Teaching. His professional experience includes writing books for students and teachers and serving as a science specialist for the Georgia Department of Education and as the coordinator of the Youth Science & Technology Center at the University of West Georgia. He is a frequent NSTA presenter and a *Science Scope* author.

Steve is a National Board Certified teacher and a recipient of the Presidential Award for Excellence in Science Teaching. He was a district director of NSTA and president of the Georgia Science Teachers Association. He is a graduate of the University of Georgia and Georgia State University. More information about the author is available at *www.sarinkbooks.com*.

Acknowledgments

Thank you to the many teachers, students, parents, and community members who helped make my outdoor classrooms a reality—in particular, my outstanding principals, Carolyn Anderson and Jeri Mansfield. I would also like to thank friends and colleagues from my professional organizations—CSSS, GSTA, NSTA, GYSTC, and SEPA—for their encouragement and friendship.

Personally, I thank those who have listened to my progress: my life partner, Glenn Bilanin; my sister, Catherine Rich Robinson; and especially my mother, June C. Rich—the exemplary nurse and outstanding educator, who nursed every injured lizard or snake I found and nurtured my love for the outdoors.

Finally, I would like to thank NSTA Press for giving my work a second life.

Index

*Page numbers in **boldface** type refer to tables or figures.*